Hydropolitics

The Itaipu Dam, Sovereignty, and the Engineering of Modern South America

Christine Folch

PRINCETON UNIVERSITY PRESS
PRINCETON AND OXFORD

Published by Princeton University Press
41 William Street, Princeton, New Jersey 08540
6 Oxford Street, Woodstock, Oxfordshire OX20 1TR

press.princeton.edu

LCCN
ISBN 9780691186597
ISBN (pbk.) 9780691186603

British Library Cataloging-in-Publication Data is available

Editorial: Fred Appel and Thalia Leaf
Production Editorial: Debbie Tegarden
Cover Design: C. G. Alvarez
Cover Art: Aerial view of Itaipu Hydroelectric Power Plant on the Paraná River along the border of Brazil and Paraguay, 14 km north of the Bridge of Friendship, connecting the cities of Foz do Iguaçu, Brazil, and Ciudad del Este, Paraguay. Courtesy of Caio Coronel / Itaipu Binacional.
Production: Erin Suydam
Publicity: Natalie Levine and Kathryn Stevens
Copyeditor: Kathleen Kageff

This book has been composed in Adobe Text Pro

Printed on acid-free paper. ∞

Printed in the United States of America

10 9 8 7 6 5 4 3 2 1

HYDROPOLITICS

Princeton Studies in Culture and Technology

Tom Boellstorff and Bill Maurer, Series Editors

This series presents innovative work that extends classic ethnographic methods and questions into areas of pressing interest in technology and economics. It explores the varied ways new technologies combine with older technologies and cultural understandings to shape novel forms of subjectivity, embodiment, knowledge, place, and community. By doing so, the series demonstrates the relevance of anthropological inquiry to emerging forms of digital culture in the broadest sense.

Sounding the Limits of Life: Essays in the Anthropology of Biology and Beyond by Stefan Helmreich with contributions from Sophia Roosth and Michele Friedner

Digital Keywords: A Vocabulary of Information Society and Culture edited by Benjamin Peters

Democracy's Infrastructure: Techno-Politics and Protest after Apartheid by Antina von Schnitzler

Everyday Sectarianism in Urban Lebanon: Infrastructures, Public Services, and Power by Joanne Randa Nucho

Disruptive Fixation: School Reform and the Pitfalls of Techno-Idealism by Christo Sims

Biomedical Odysseys: Fetal Cell Experiments from Cyberspace to China by Priscilla Song

Watch Me Play: Twitch and the Rise of Game Live Streaming by T. L. Taylor

Chasing Innovation: Making Entrepreneurial Citizens in Modern India by Lilly Irani

The Future of Immortality: Remaking Life and Death in Contemporary Russia by Anya Bernstein

Hydropolitics: The Itaipu Dam, Sovereignty, and the Engineering of Modern South America by Christine Folch

To Daniel

FIGURE 0.1. Aerial photograph of Itaipú Binational Dam and Reservoir.
Source: Author's photograph.

CONTENTS

ILLUSTRATIONS

Figures

Tables

ACKNOWLEDGMENTS

A book like this entails numerous debts of gratitude, which I have happily accumulated over the years, from the start of the book as the germ of a question raised during a grad school seminar to the finished manuscript. The acquisition of more debts than I can enumerate has been a great joy, and the inadequacies in this manuscript are my own and not those of the readers, reviewers, and interlocutors with whom I have had the fortune to engage. I first thank Princeton University Press and editor Fred Appel for their careful stewardship of this book as well as series editors Tom Boellstorff and Bill Maurer for their support of the manuscript even before it was fully drafted. Two anonymous reviewers gave generous, rigorous, and extensive feedback on several versions of the manuscript and have, to my thinking, established a high bar for peer review, which I hope to pay forward.

The research that forms the original core of this book began as a dissertation project at the CUNY Graduate Center, where I was in the PhD program in cultural anthropology. I thank my exam and dissertation committee members who helped me stake out a set of questions ranging from political anthropology to political ecology: Marc Edelman, my dissertation chair; Fernando Coronil; David Harvey; Neil Smith, whose encouragement to pursue a budding interest in Paraguay led to my first trip to the Southern Cone; and Katherine Verdery, who regularly challenged me on how to conceptualize the state. Faculty in the program asked probing questions and helped me develop the project; I thank Kevin Birth, John Collins, Vincent Crapanzano, Gerald Creed, Louise Lennihan, and Don Robotham; and Julie Skurski for her careful reading of many chapters. Among my peers who were my most important interlocutors while in graduate school, I especially thank Alessandro Angelini, Fadi Bardawil, Chris Caruso, Adrienne Lotson, Cam McDonald, Slobodan Mitrovic, Nada Moumtaz, Theodore Powers, Jeremy Rayner, and Melissa Zavala. I received dissertation fieldwork support from the Wenner-Gren Foundation and the Fulbright IIE in order to conduct research in Paraguay and the Triple Frontier. I wrote the dissertation with support from an American Council of Learned Societies–Mellon Dissertation Write-Up Grant and as a member of the New York Public Library's Wertheim Study, a haven for writers and researchers across the city at that time ably coordinated by librarian Jay Barksdale.

My debts in Paraguay are even greater as individuals from all walks of life and social commitments shared their wisdom and their networks and their strong opinions about how history and politics and hope for the future were tied up in Itaipú. I first thank the thriving anthropology community centered between the Catholic University of Asunción and the Museo Etnografico Andres Barbero, especially Marilín Rehnfeldt, who heads up the graduate anthropology program at the Catholic University, and the Museo's director, Adelina Pusineri, and vice director, Raquel Zalazar. Not only did the Museo offer me affiliation while I was on the field; it hosted the first event where I shared research findings in Paraguay. The energy systems researchers of the National University of Asunción–Polytechnic Grupo de Investigación en Sistemas Energéticos (GISE) have become critical conversation partners in recent years and have firmly placed Paraguay-based energy research on the leading edge of energy policy investigation internationally. I am especially grateful to GISE director and founder Gerardo Blanco, Cecilia Llamosas, and the National University of Asunción–Polytechnic research director Victorio Oxilia, as well as to UK-based legal scholar Maria Gwynn. Though it's impossible to name all the Paraguayans with whom I spoke on Itaipú, I highlight Ricardo Canese and the social movement Tekojoja, who were core members of the Lugo government's renegotiation team, and Carlos Mateo Balmelli, former executive director of Itaipú, and the energy experts of the Itaipú Executive Directorate. Crucial archival evidence for this project came from the Center for Documentation and Archive for Defense of Human Rights (Asunción); I thank director Agustín Fernandez and coordinator Rosa Palau. While in Paraguay, I met key researchers who have become indispensable scholarly interlocutors on the landlocked country: Lucas Arce, Kregg Hetherington, John Tofik Karam, Caroline Schuster, and Gustavo Setrini. For their friendship in Paraguay, I also thank the Saint Andrews community, Hermelinda Aguayo, Edu Barreto, Osvaldo Bittar, Tania Buzarquis, Juan Carlos Cristaldo, Nahum Dam, Cristina del Puerto, Ivan Evreinoff, Gabriela Galilea, Monica Galilea, Raul Gutierrez, Richard Lavieille, Selva Morel, Edson Torres and his family, Federico Torres, and Ampelio Villalba; and Luis Maria Duarte, whose life was tragically cut short while defending transparent elections in Afghanistan. Above all, I have been inspired by the young people of Paraguay, who have asked the most probing questions and have a vision for transformation for their country.

Colleagues and friends and other fellow travelers have played crucial roles. Friends from college and former roommates: Agbenyo Aheto, Solita Alexander, Abena Osseo Asare, Ana Balcarcel, Nathaniel Barksdale, Nathan Becker, Elliot DeHaan, Katy Hsiao Goldsborough, Dan Horwitz, Sila Jamalludin, Olamide Jarrett, Liza Cagua Koo, Jonathan Liu, Robyn Runft Liu, Megan White Mukuria, Dakota Pippins, Alicia Ingalls Pittard, John Pittard, Rebecca Ray, Lydia Johnson Reynolds, Jane Rubio, Kenji Scott, Naunihal Singh, Jae Suh, Matthew Suh, Tina Teng, Julie Thwing, Kaniaru Wacieni, Ming Wei, Yong Yeow Yeoh, and Margaret Yoon. Writers, artists, and intellectuals from New York and Chicago opened crucial conversations: Kwame Akowuah, Christina Winters Blaustein, Garnette Cadogan, Lucia Cantero, Teju Cole, Bob Goldsborough, Parneshia Jones, Jan Kang, Gretchen Marsh, Luz Posadas, Julie Rodgers, Manuel Schwab, Kirsten Wieser Scott, Li Lian Tan, Brie Walker, Carey Wallace, and the Chicago book club. At Wheaton College, I was part of a vibrant scholarly community with excellent colleagues Hank Allen, Gene Green, Brian Howell, Ezer Kang, Jan Kang, Henry Kim, Vanya Koo, Michael Mangis, Brian Miller, Leah Samuelson, Noah Toly, Laura Yoder, and the Whisk(e)y Society; as well as with undergraduates who provided critical feedback on chapters, including my research assistants Michael Daugherty and Alyssa Overturf and the students of the Anthropology of Energy seminar. I am especially grateful for the faculty reading group I was a part of; we exchanged book chapters, article drafts, and invaluable advice: Leah Anderson, Larycia Hawkins, Sandra Joireman, Amy Reynolds, Christa Tooley, and Rachel Vanderhill.

Duke University cultural anthropologists of Latin America have read full manuscripts and helped me work through ideas: Diane Nelson, Irene Silverblatt, and Orin Starn. And from my very first visit to the university, my home department has provided keen insight (and warm collegiality): Anne Allison, Lee Baker, Engseng Ho, Ralph Litzinger, Anne-Maria Makhulu, J. Lorand Matory, Laurie McIntosh, Louise Meintjes, Charles Piot, Harris Soloman, Rebecca Stein, and Charles Thompson. Thanks also go to the vibrant community of scholars of Latin America gathered at Duke and beyond: Global Brazil Humanities Lab codirectors John French and Esther Gabara; and Paul Baker, Brandon Bayne, Jake Blanc, Arturo Escobar, Frederico Freitas, Gustavo Furtado, Ellen McLarney, William Pan, Cynthia Radding, Ildo Sauer, Elizabeth Shapiro-Garza. Duke is strongly committed to interdisciplinarity, and faculty from across the university have become delightful conversation partners: Luke Bretherton, Christena Cleveland, Mona Hassan, Meredith Riedel, and Norman Wirzba as well as the Duke Energy Initiative staff and director Brian Murray. Because of support from a Duke University Franklin Humanities Institute (FHI) Faculty Manuscript Workshop Award, Duke colleagues provided important insights on a later draft: among others, Michaeline Crichlow, Jocelyn Olcott, Dalia Patiño-Echeverri, and FHI's then-director Deborah Jenson, who, along with FHI associate director Chris Chia, was crucial in making the institute an intellectual home during my years codirecting the Global Brazil Humanities Lab. Barbara Rose Johnston and Kregg Hetherington traveled great distances to join the manuscript workshop. The Duke Center for Latin American and Caribbean Studies (CLACS) has been a partner in my research from my graduate school days. The center provided early research support in the form of a Foreign Language and Area Studies (FLAS) scholarship to study Guaraní while I was a graduate student at CUNY. Now as Duke faculty, we have continued the relationship, and I thank director Patrick Duddy, associate director Natalie Hartman, Antonio Arce, Ken Maffitt, and Miguel Rojas-Sotelo. I am grateful to CLACS for the Junior Faculty Book Award granted for this monograph and for the University of Wisconsin Cartography Lab for work on graphs and maps in the manuscript. Duke students, undergraduate and graduate, have provided enthusiastic engagement on renewable energy politics in Latin America—my research assistant Olivia Sanchez, along with Emily Davenport, Odette Rouvet, and Connor Vasu—and have taught me about public-facing scholarship. I also thank the Emmaus Way community, deeply committed to spiritual practice and lived justice, and especially Pub Group for the salubrious discussions on liberation theology, critical race theory, and poetry.

Love of learning, music, and curiosity were values instilled from the very beginning by my family. My Cuban grandparents, Antonio and Ondina, and my Dominican grandmother, Germania, were lovers of music and astute readers of scripture, and, though they are not here to see this book, the lessons they taught and the stories of their lives quickened my interest in the region. My parents, Juan and Julie, taught me to love books and were so committed to my intellectual journey and curious about my research that they even joined me in Paraguay during my fieldwork, where we wandered around the country and visited the waterfalls of Iguazú—they especially loved Encarnación. My aunts, uncles, and cousins William, Wendy, Ingrid, Emilio, Soraya, Wesley, Jenny, Caroline, Fernando, Ricky, Tony, Soly, Alex,

Christian, and Desiree have been a constant support. My new family in Paraguay has brought unending *asados*, late-night conversations, and delight: Diana and Victor offer great insight on all things Paraguay and are wonderful grandparents; Victor, Mercedes, Soledad, Emilia, Liliana, and Daniel's two big sisters, Juani and Mica, who adore him and whom he loves. My siblings have been my coconspirators all my life, and it is a joy to see how our families have grown: Marcus and Celeste, Candace and Jakob, and a crew of nieces and nephews: Andromeda, Elias, Mattin, and Emmeline.

And most of all, I thank Juan Carlos for adventuring with me in this life, and for his profound analysis of politics, the dialectic, and the history of Paraguay, and I thank Daniel for unexpected joys every day. You both have surprised me, and I love you more than I can say.

HYDROPOLITICS

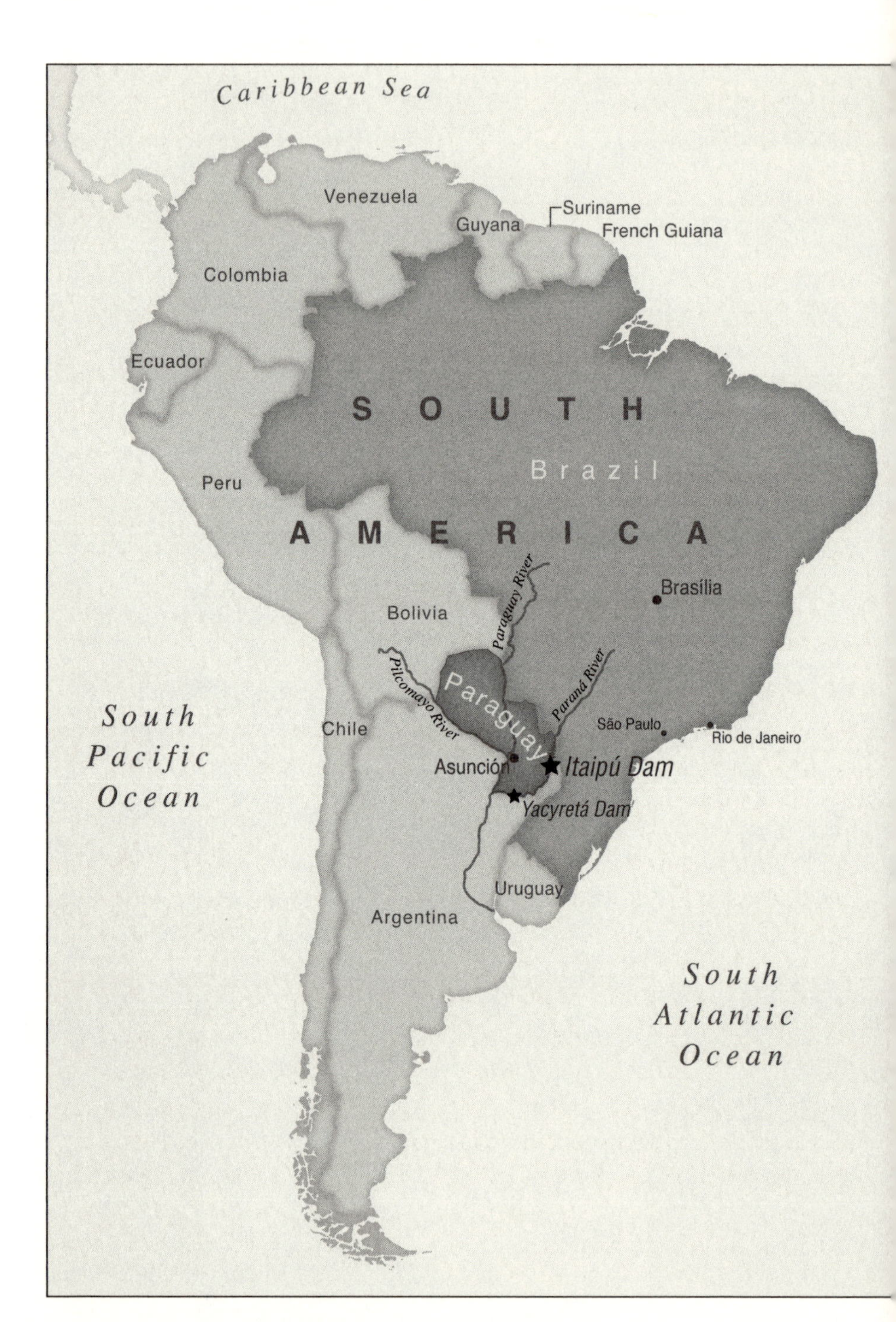

FIGURE 0.2. Map of South America.
Source: Designed by the University of Wisconsin Cartography Lab

Introduction

Asunción, November 2008

In the face of an energy crisis with a straightforward technical solution on the table, the managers of the world's largest dam decided instead to cross their fingers and hope for rain. That it worked was more than a stroke of good fortune but rather illustrates how infrastructure engineers political power.

"They are offering to open the gates of the dams upriver so that the reservoir fills up," explained Carlos Mateo Balmelli to me as we sat in the fourth-floor corner office of Itaipú Binational Hydroelectric Dam's (Brazil-Paraguay) Paraguayan headquarters.

The dry period of 2008 finally came to a breaking point at the beginning of the famously long and sweltering Paraguayan and Brazilian summers. Balmelli, the Paraguayan executive director of Itaipú Dam, fielded multiple phone calls and frenzied office visits from engineers who exchanged hydrological reports and computerized modeling of expected precipitation levels. After months of meager rains, the Paraná River was low, yet some in Itaipú's binational leadership wanted to take advantage of increased projected demand and produce more electricity than warranted by the current amount of water in the reservoir. But without new rains, the increased production would push the Itaipú reservoir dangerously close to its minimum fill level, and, as demand for electricity increased during the summer months in Brazil and Paraguay, the shortfall would leave the dam unable to meet the need.

The problem seemed to have an obvious solution: if some of the more than forty dams farther north on the Paraná were opened (all of which lay squarely within Brazil's national territory), their gates could send more water downstream to the Itaipú reservoir, and the dam would have enough to meet production needs. And because Itaipú was more efficient than the forty upriver dams, more energy could be generated by the same quantity of water.

But disapproval was unmistakable in Balmelli's voice.

"Why not let upriver dams open their gates to allow more water into the Itaipú reservoir?" I asked.

"Sovereignty," he answered, not skipping a beat. After a moment he added, "Letting Brazil control the water flow to help Itaipú meet its production makes Paraguayan sovereignty dependent and subordinate to Brazil."

Maintaining the landlocked country's independence from its largest neighbor was an ever-present anxiety in Paraguayan public discourse, with roots in the devastating War of the Triple Alliance (1864–70), as we will see. And so, the Paraguayan half of the Itaipú Dam Executive Directorate Board—wielding their veto power, engineering reports, and the weather forecast—insisted that energy production be increased with no other action taken. Rain came, and Itaipú shattered world records for energy production that year.

Administration of the largest hydroelectric dam in the world required careful behind-the-scenes calibration, not just of the rate at which generating units spin or of the quantity of water passing through the turbines, but of national sovereignty. Within social science, the notion of national sovereignty has multiple working definitions, including: the ability to ward off outside intervention, a vertical ability to decide and impose a state of exception on a population, a horizontal recognition by other sovereigns as the legitimate political authority within a geographic unit; and in international law sovereignty doctrine is particularly concerned with control over desirable natural resources.[1] All these understandings aptly apply to some aspect of governance in Paraguay in general and to Itaipú Hydroelectric Dam in particular. Because sovereignty is conceptually connected to territory, the condition of landlockedness often heightens anxieties all the more. But in the case of the choice to wait for rain, the Paraguayan managers of Itaipú Dam reimagined sovereignty as the ability for the dam to work independently of the Brazilian hydroelectric system. That is, sovereignty had to do with the self-provisioning of the nation via hydrological processes outside of human control.

Within Paraguay, the engineering decision sparked a political scandal that made news headlines and even saw Balmelli called into the presidency to give an account of the Paraguayan Executive Directorate's actions. Summoned by the president and the foreign minister to justify the decision to increase Itaipú production before the end-of-year rains, Balmelli played the ace up his sleeve. In addition to documenting the technical feasibility and that previous administrations had done the same thing to no detriment, in addition to showing how much revenue the Paraguayan treasury had received, and in addition to producing written communication between the Paraguayan Executive Directorate and the foreign minister himself months before about the need to "flexibilize" Itaipú production, he pointed out that the major consumer of the extra electricity was Paraguay, not Brazil.

Had Itaipú relied on the upriver dams to make the additional production viable, Paraguay's ability to supply itself with badly needed electricity would have depended on the generous goodwill of its neighbor.[2] For any other Paraguayan administration, that might have been galling. But progressive Paraguayan president Fernando Lugo (2008–12) had unseated the six-decade-ruling Colorado Party by decrying previous governments as too beholden to Brazilian interests in Itaipú amid an unprecedented hemisphere-wide swing to the left. His government had pledged to "recover Paraguay's hydroelectric sovereignty" in the binational dam—a rallying cry that had captured the imagination of voters weary of business as usual, exemplifying the prominent role of natural resources in Latin America's leftward political turn of the early twenty-first century. Depending on the generosity of Brazil would have been disastrous, especially since the two administrations were embroiled in another round of tense negotiations over Itaipú at the time. The other energy option would have been unplanned and therefore more expensive electricity from Paraguay's other binational dam, Yacyretá (Argentina-Paraguay), which may have led to politically unpopular electricity rationing. The president and foreign minister were mollified by the line of reasoning that balanced technical feasibility and political expedience, and the matter was settled. The Paraguayan Executive Directorate landed on a possible, if not strictly optimal, solution that protected the administration's political stature.

For the Paraguayan energy managers and politicians beholden to an electorate that had supported Lugo, the electricity of the dam was not just energy or just a source of revenue, but a sociopolitical substance that could express sovereignty. In Itaipú, engineering solves state problems by making and shaping environmental interventions to such a degree that it has become a way of *doing politics*. Politics may be studied in law or in the tensions between the textual and the actual; it may be studied in the actions of state actors or those on its contentious margins; it may be studied in ideology, ritual, or institution. But by denominating the subject of this book as "hydropolitics," I follow the political via water. Hydropolitics, the political economy that comes from an industrialization and electrification powered by water, acknowledges how it comes to matter for politics in all the definitions above that hydro is the source for power. Particularly, I uncover how it comes to matter to politics within Paraguay that the electricity that powers homes and factories comes from hydro, not fossil fuels.

To untangle how energy can be simultaneously technological and sociopolitical, this book proceeds from a simple premise: our relationship to the environment is a form of cultural production, which, in turn, inflects political, economic, and social structures. That is, narratives, values, and aspirations mark and emanate from electricity. Understanding the dam requires the dual intervention of political ecology, which analyzes both how human interventions shape environment and how the shaping of nature in turn affects human communities. Itaipú has presented the Brazilian and Paraguayan governments the ability to achieve multiple political goals and has had far-reaching cascade effects, intended and unintended. What Itaipú Dam has done is to turn the Paraná River under its influence into a political-electrical machine, an engineered complex of geological objects, atmospheric cycles, and cement intrusions.[3] But these are densely packed claims. The next few pages will use a series of historical and ethnographic moments centered on the role of Itaipú in Paraguay to show how energy is cultural production, the historical context that makes Itaipú engineering both political and electrical, and the place of electricity in the new political-economic options of Latin America's left turn.

Energy as Cultural Production

Itaipú Binational Hydroelectric Dam, the world's largest producer of renewable energy, sits across the Paraná River, which divides Paraguay from Brazil in the heart of South America. The two governments that own the dam like to boast that the power plant is one of the seven wonders of the modern world. Praises aside, it was not just government elites in closed-door rooms who had strong opinions about how Paraguayan sovereignty was tied up in Itaipú electricity. The popular dissatisfaction so instrumental in putting Fernando Lugo into office was evident from the very first time I visited the

Brazilian-Paraguayan dam as part of a tour group from Asunción, Paraguay, vacationing in Foz do Iguaçu, Brazil, for the April 2007 Semana Santa holidays.

Along with forty other visitors from Paraguay, I dutifully followed the Brazilian tour guide to an observation deck overlooking the placid river. Using comparisons like "the amount of iron and steel in 380 Eiffel Towers" or "210 Maracanã Stadiums worth of concrete," he attempted to explain the grandeur of Itaipú Binational Dam. But I still had my recent visit to Iguazú Falls on the mind, waterfalls so photogenic they had been featured in Hollywood films from *Moonraker* to *The Mission* to the fourth Indiana Jones installment. The impression left by the famous cataracts on the Iguazú River border of Argentina and Brazil, waters rushing, pounding rock so forcefully that the mist looked like smoke, eclipsed the bland concrete wall of the hydroelectric dam and calm reservoir set against an overcast gray sky.

The current in the Paraná River was barely noticeable. Not even the spillways were open. In fact, it looked like nothing at all was happening in the "mightiest dam in the world." While I only half listened because the data felt contextless, it did strike me as odd that a bus full of Paraguayans would have gone to the Brazilian side of the dam for a tour.

And then, one Asunceno to my right leaned in and muttered underneath his breath, "Paraguay used to have waterfalls like Iguazú."

He must have taken my look of surprise as invitation to go on.

"But they were destroyed for that," he added with a meaningful nod at Itaipú Dam. The tour guide continued blithely on, never mentioning the Guairá Falls, which once lay 150 miles upstream of Itaipú Dam. Like Iguazú Falls on the eponymous river, the Guairá Cataracts on the Paraná River had been carved by flowing water into the rocky terrain that dominated the triborder region between Argentina, Brazil, and Paraguay. But unlike Iguazú, which was a major tourist destination for both Argentina and Brazil, it was now impossible to visit Guairá Falls. They were permanently submerged under the waters of the Paraná River when the Itaipú Dam reservoir was filled in 1982.

From the observation deck, it was hard to appreciate the scale of the dam, but once I finally donned a hard hat for a technical tour a year later, I found that, as with the Iguazú Falls, the photos and distant vistas of Itaipú Dam do not do it justice. The dam's five-mile-long walls, made of rock fill, earth fill, and concrete, form an artificial lake 105 miles in length, possessing a width so great that the other bank is barely visible from some locations. The vertigo-inducing main structure rises 643 feet, about sixty-five stories, where twenty white penstock tubes—each with a diameter of thirty-four feet (bigger than giant sequoias)—recline against gray concrete. Below, in the generation gallery, engineers ride go-carts to cross the distance. Underneath the floor of the gallery, water that has rushed through the penstocks rotates turbines ninety-plus times a minute to generate world-record-shattering amounts of electricity. Because of the way it is produced, hydroelectricity cannot be stored. At times when energy consumption is expected to increase, Itaipú's intake system allows more water to enter the turbines so that more electricity is generated. A workforce of engineers and technicians is needed twenty-four hours a day to monitor the energy production, calculate estimated demand, and maintain the physical dam, even with the advent of computer technology.

When Itaipú is at optimal efficiency, it looks like nothing much is happening from the outside. The reservoir is tranquil, and the river below the dam has a steady but not turbulent current. Three or four times a year, however, the water level in the Paraná rises to such a degree that the engineers have to open the thunderous spillways to drain the excess. In my years of visiting the dam, I have seen the spillways open only once, their white rapids recalling the majestic Iguazú Falls just a few miles away. Infrastructure of this kind—simultaneously monumental and practical—is not human scaled. Tour guides use comparisons with the Eiffel Tower and the Maracanã Stadium to give visitors a sense of the soaring dimensions. But because the principal output of the dam lies at the subatomic level, the flow of electric charge, Itaipú also rests at a scale too small for human eyes to see.

Through the meticulous monitoring of water levels, temperature, and flow, Itaipú integrates the Paraná River basin from bedrock to cloud cover. Because of regular communication with the upriver Brazilian dams and Yacyretá Dam downriver, almost the entire length of the Paraná has become a coordinated system. What looks natural is a kind of inorganic cyborg, a natural machine that depends on the cyclical availability of nonhuman processes. Itaipú represents a significant energy infrastructural investment. It was incredibly expensive to build; with a rated life span of 150 years, it is long-lived; and it is practically irreversible—thus creating political, economic, and social path dependencies that mean that, even were Itaipú somehow removed today, Brazil and Paraguay would still have to contend with dynamics put in place by the dam. From water molecules rushing through turbines to electrons moving through wires, from the spreadsheets of engineers to the budget balances of politicians, and from factory conveyer belts to a family cooking dinner, this book is organized around the theme of energy conversions. Through the physical infrastructure, the binational political arrangement, and the financial liquidity of energy monies, Itaipú Dam converts water into circulations, integrations, and futures.

As we have increasingly become aware of the threats of global warming and the deadline imposed by peak oil, the implications of energy sourcing become all the more obvious and politically fraught. A statistic may help illustrate the point: the majority of electricity generated in the United States—two-thirds—comes by burning fossil fuels.[4] And this primary dependence on fossil fuels runs across the globe: Europe, Africa, and Asia derive two-thirds to three-quarters of their electricity from coal, oil, and natural gas; and in the Middle East, nearly all electricity is derived from fossil fuels. The sole outlier to the trend is South America. Nearly two-thirds of the electricity produced in South America is from renewable energy sources, chiefly hydroelectric dams. Renewable sources are not "alternative" sources in the region; rather, hydropower is to politics, economics, and society in South America what coal, oil, and natural gas are in the United States, offering a glimpse into what a post-fossil-fuel future might look like.

Amid growing demands for energy and the challenge of anthropogenic climate change, the renewable energy sources of the twentieth century are likely to become the mainstream energy sources of the twenty-first outside of just South America. But the role of hydropower is even more curious; some of the world's largest confirmed hydrocarbon deposits are in the region. Petroleum exporter Venezuela, for example, gets 70 percent of its electricity from hydropower; Ecuador gets 49 percent; Colombia gets 70 percent; Peru gets 49 percent but has invested in expanding solar and eolic significantly since 2012.[5] Paraguay gets 100 percent of its electricity from its dams, and regional giant Brazil gets 70 percent from hydroelectric sources, nearly 20 percent from Itaipú alone. So, then, why would a region famous for the exportation of fossil fuels be so internally reliant on hydropower instead? What political and economic opportunities and challenges are opened up when the foundation of an economy is hydroelectricity? And what difference does that make? Itaipú illustrates the challenges of providing the carbon-neutral energy required to meet the needs of the planet.

The energy infrastructure of the dam not only powers the daily lives of millions as a seamless part of the customs and aspirations of Brazilians and Paraguayans. It also gets drawn into meanings that connect to the broader worldviews of citizens of both countries. In this book, I show how and why sovereignty and security dominate energy politics in Itaipú and the region beyond as engineers use the dam to convert differences in height into electricity and as its managers convert energy into political-economic goals. The dam's ability to produce energy, sovereignty, and security constitutes a set of competing priorities that govern the political processes of infrastructure and integration. Consequently, Itaipú has resulted in the formation of a new kind of territory contoured along electrical wires and waterways (and not merely nation-state borders) and in such tight circulations of electrical charge and money that the distinctions between the two blur, suggestive of something more like an ecocurrency. These political dynamics arise from the base of the dam: water.

A long-standing body of literature has queried the role of water in human history. Karl Wittfogel introduced the idea of "hydraulic civilizations" in his account of the lynchpin function of irrigation in the formation and centralization of early agricultural state societies.[6] This is not a quaint archaeohistorical debate; the crises of the Anthropocene (the current geologic epoch so named in recognition of human activity as an earth-shaping force) have resulted in a new negotiation of how human society engages with the environment both in scholarship and in policy.[7] Whereas Wittfogel dealt with water management in ancient societies, I am interested in energy in advanced capitalist societies. South America's dependence on renewable energy as the primary source of power for industrialization has resulted in a region deeply impacted by the material constraints of hydroelectricity. Because of the transboundary geographic distribution of major waterways in South America and the fact that water is necessary for life (fossil fuels have few uses other than energy generation), hydroelectric resources have heightened significance. Itaipú Dam's infrastructure has demanded an unprecedented level of physical and political cooperation between Brazil and Paraguay as well as Argentina, Bolivia, and Uruguay, an integration that has redefined national sovereignties.

Framing energy as cultural production highlights how it is indissolubly both physics and politics, the result of mechanical requirements and social context. In anthropology, this is actually a very old argument. Leslie White claimed that energy was the limiting factor in social development and that intensification of energy access—for example, controlling fire, then switching to the steam engine, coal furnaces, and eventually nuclear fission—was directly responsible for increased societal complexity.[8] Moreover, nearly a century before White's groundbreaking article, early anthropologist Lewis Henry Morgan asserted that, *pace* eugenicist arguments of his day, the difference between different people groups was not biological superiority and inferiority, but instead the environment.[9] Although the technocentric claims of Morgan, White, and Wittfogel fell out of popularity with the poststructural turn and a warranted skepticism toward overly deterministic arguments, ecological crisis has insistently foregrounded these questions anew.[10] Weather is famously fickle, but climate change has added greater uncertainty to hydroelectricity.

Hydroelectric Histories: Making a Political-Electrical Machine

Like many tales told in Paraguay, the story of Itaipú Dam takes its departure from the War of the Triple Alliance (1864–70). How two governments sacrificed the world-historic natural beauty of Guairá Falls for the sake of energy extraction begins in semantics and a border dispute. In the 1870s, Paraguay finally signed separate peace accords with the occupying forces of Argentina and Brazil after the disastrous war, which it fought and lost against the combined armies of Argentina, Brazil, and Uruguay. With nine out of ten Paraguayan males (including children) dead in the fighting and only a fraction of the women remaining alive, the devastation left the Paraguayan government in shambles and the entire territory sparsely populated. It is impossible to overstate the significance of the war in Paraguayan memory and identity making, and most of my Paraguayan interlocutors began their stories about the dam with a reference to the "Great War."[11]

Treaties assigned a third of the Paraguayan national territory in the east to Argentina and Brazil and fixed new borders for the three countries. But the 1872 Brazil-Paraguay peace treaty ensured future conflict with the use of single word: *hasta*. *Hasta* is a preposition often translated as "until" or "up to" but may or may not also mean "including." The 1872 treaty stated:

> The territory of the Empire of Brazil is separated from that of the Republic of Paraguay by the stream or riverbed of the Paraná River, where Brazilian possessions begin at the mouth of Iguazú up to [*hasta*] the Great Fall of Sete Quedas of the same Paraná River. From the Great Fall of Sete Quedas, the dividing line continues by way of the Mbaracayú mountain range until it concludes. (article 1)

Guairá Falls[12] (or Sete Quedas) was a series of steep cataracts breaking on the Paraná River between Paraguay and Brazil as the river wound through a narrow canyon east of the Mbaracayú mountain range. Indeed, the falls possessed a flow rate greater than any other waterfall on the planet. The heavily forested terrain, far from Brazilian or Paraguayan metropoles, offered some of the last relatively untouched land where the nomadic Guaraní peoples who had once dominated the landscape from southern Bolivia to eastern Brazil might live in their traditional communities. For decades following the end of the war, governments in Brazil squabbled with governments in Paraguay over the precise location of the border between the two countries. Brazil argued that its boundary with Paraguay "included" and "encompassed" the falls whereas Paraguay argued that it "went up to" the falls and thus they were shared by both countries.[13] In the 1950s, engineers in Brazil began discussions in earnest about how to take advantage of the hydroelectric potential of the Paraná River in the southwest of Brazil. There was even talk of selling energy produced by Brazil to Paraguay, to compensate for any effects on the Paraná's flow that upriver dam building might have.

As energy experts in Brazil sketched plans for the river, public posturing in Paraguay—where memory of the war and anxieties expressed in terms of the country's landlockedness took their form in a discourse of fierce patriotism—reached a fevered pitch.[14] To halt Paraguayan claims on the waterfalls and on their touristic and hydroelectric potential, the Brazilian military government moved a small number of troops near the falls in June 1965. War loomed. But Paraguay had no interest in a war it would lose. The military government in Paraguay reacted to the casus belli with delicacy, employing a tone of dialogue and judicious diplomatic complaint with Brazil and putting down the growing public protest within Paraguay with force. As time passed, the conflict garnered international attention, mostly negative toward Brazil. Finally, in June 1966 Brazil's foreign minister Juracy Magalhães invited his Paraguayan counterpart Raúl Sapena Pastor to meetings in Foz do Iguaçu, Brazil, to attempt to find a resolution to the matter.

Whereas the Paraguayan team consisted of other members of the Ministry of Foreign Relations (including the Paraguayan ambassador to the United States), top members of the ruling Colorado Party, and members of the Border Definition Commission, the Brazilian team included the minister of mines and energy and the president of Furnas (a subsidiary of Eletrobras), who was an engineering expert on hydroelectricity. The difference between the two teams reflected a difference in their priorities. For the Paraguayan side, it was a matter of attempting to define the border and of redeploying the Brazilian troops; for the Brazilian side, it was a matter of energy. Within international law, the borders between two sovereigns can be redefined only through common agreement (but in Brazil there was no will to change) or war. At the unresolvable impasse over possession of Guairá Falls, the Brazilian foreign minister made a provocative argument that won the day: Magalhães suggested submerging the entire area of dispute in a reservoir for a future dam on the Paraná.

The two foreign ministers signed the 1966 Act of Foz do Iguaçu, agreeing to build and share a dam on the Paraná. Although nationalist sentiments ran high, the 1960s and 1970s saw strengthened south-south relations worldwide as a means of self-rule. Years of preliminary studies followed, and, in 1973, the Itaipú Treaty, creating Itaipú Binational was signed and passed with great fanfare, a sign of the global trend toward cooperation between developing and/or nonaligned countries. To construct the dam, a total workforce of thirty-five thousand labored from 1975 to 1991 in three overlapping phases of construction marked by civil engineering, then mechanical engineering, and lastly electrical engineering. Complete minicities, containing homes, schools, churches, movie theaters, and even a brothel, were built in both Brazil and Paraguay to house the laborers and their families.[15] One hundred forty-nine workers perished in the entire endeavor, a sacrifice to modernize both countries. The Ava Guaraní lost their ancestral lands on the riverbanks, another sacrifice, and though the workers who died are memorized in an orchard of carefully planted trees, there is notable silence around Native realities. Some of the thirty indigenous communities living on the Paraná were relocated to worse lands; many others were not indemnified. Instead, the Natural History Museum (later rebranded and replaced by the Land of the Guaraní People Museum, Museo de la Tierra Guaraní) opened in 1979 on the Paraguayan side of the dam close to the site of the Itaipú zoo. The museum displays archeological artifacts encountered during the civil works and shows how nineteenth-century visitors to the region depicted indigenous daily life, relegating indigeneity to the past.

During the civil works phrase, workers removed earth and rock to build a diversion channel (1975–76) next to the planned location of the dam to eventually drain the riverbed. Once the finished diversion channel was dynamited open in 1978, temporary dams were built just north and just south of the site for the main structure, allowing the Paraná to take a different track while the original path was drained. When the riverbed finally dried in 1979, the crew, working twenty-four hours a day, built a cathedral-like hollow dam with concrete rated to last 150 years. The main structure was finally complete in 1982, when the reservoir could at last be filled. Over the next decade, as mechanical engineering (the building of machines) came to the fore, specially designed turbines were brought to the dam; the transportation required to move them slowly across Brazil's roadways prematurely aged and depreciated the asphalt. As each turbine was finally installed (beginning in 1984 and ending in 1991), it was brought online, linking mechanical to electrical engineering. Electrical substations, one for ANDE (Administración Nacional de Electricidad) and one for Eletrobras, were simultaneously built and expanded, ready to transport the newly available energy to growing markets.

The sheer size and technical efficiency of Itaipú established engineering standards for four decades. Itaipú's twenty turbines have an installed capacity of fourteen thousand megawatts, capable of singlehandedly powering 20 percent of Brazil's total electricity needs, seven Paraguays, or the equivalent of about a third of the state of California's consumption.[16] Because of its much smaller demand, Paraguay consumes only a fraction of its half of the electricity from the dam (which supplies about 85 percent of its electricity). It exports the majority of its power to its neighbor for a controversial, below-market price. In actuality, Itaipú produces anywhere between 17 to 19 percent of Brazil's annual demand.[17] Most of the dam's nearly US$4 billion in energy sales every year comes from Brazilian consumers.

Itaipú was created to resolve tensions by internalizing and thus harmonizing competing agendas, but what occurred was a rescaling of the border disagreements and sovereignty claims. In international law, Itaipú was sui generis, the single example of its kind, legally unprecedented, and without any sort of custom or pattern to follow. Yet the binational diplomatic cooperation between Brazil and Paraguay over Itaipú put in motion a series of treaties and international agreements within the entire region that resulted in the 1991 creation of a local equivalent of the European Union—the Common Market of the Southern Cone (Mercosur)—by founding members Argentina, Brazil, Paraguay, and Uruguay. The dam functioned within and beyond the existing Brazilian and Paraguayan state structures because it was designed as a juridically distinct binational entity, a space of exception outside the jurisdiction of either country, ostensibly to protect the sovereignty of each party. In practice, of course, the relationship of institutions, individuals, and resources could never be so neatly assigned, leading to state-sponsored violence, resource management to benefit certain groups and not others, and a "mask" of justification via narratives of development and sovereignty.[18] The entire leadership and workforce was divided between the two countries, requiring a careful and contentious balancing act of cooperation versus independence in the face of vast inequalities as the dam has served as an international diplomatic organization since its beginning. To further complicate matters, today the dam is steered by two kinds of senior employees with distinct training and objectives: political appointees (like Balmelli) and technicians (who interpreted the hydrological reports to him).

As with electricity and law, so with money: Itaipú required a level of financing hitherto unseen, with economic practices whose complications continue into the present and patently reflect the asymmetries between Paraguay and Brazil. The world's largest energy producer had correspondingly high expenses *and* sales. The debt-financed construction—most of which was loaned by the Brazilian government itself—was at the mercy of fluctuations in global credit markets and commodity prices strained by the oil crises of the 1970s, the economic instabilities of Latin America's "lost" decade of the 1980s, and the expected graft and embezzlement attendant with megaprojects. However, the dam's controversial debt grew from a planned grand total of US$2 billion (as estimated in 1972) to US$16 billion (in 1984) to US$60 billion (today) through more than just fiscal malfeasance or economic downturn, but rather via financial planning that depended on and then extended the reach of the state (chapter 4).

Promises of neighborly equity notwithstanding, financial inequality and sour grapes characterize Itaipú energy sales (chapter 2). Though the two governments each own 50 percent of the energy potential in Itaipú's twenty turbines, the distribution of that energy has reflected the difference in demand in Paraguay and Brazil. From 1985 to the present, Paraguay's public utility ANDE contracted 7 percent of the potential of the dam to sell to the Paraguayan market; the Brazilian utility Eletrobras contracted the remaining 93 percent. A substantial portion of the 93 percent, obviously, has come from Paraguay-owned turbines. For this "ceded" energy, Eletrobras "compensates" the Paraguayan government an amount that is currently about US$9 per megawatt hour (or around US$360 million a year). The Brazilian utility then resells that energy for anywhere between US$20 and US$60 more per megawatt hour. Per the 1973 treaty, Itaipú operates in the neutral currency of US dollars. Nevertheless, the finances of Itaipú—particularly the amount paid to Paraguay by Brazil for ceded energy and the enormous debt that burgeoned as a result of underpayment, compound interest, and conflicts of interest—are a sore topic in both countries.

The binational arrangement shifted once more when, from 2008 to 2012, Paraguay underwent leftward regime change. The new government attempted to renegotiate the power-sharing relationship with Brazil in the dam and to change how energy rents were spent within Paraguay, even going so far as to invent the new political category of "hydroelectric sovereignty" to describe the aims and scope of sovereignty.[19] For Paraguayan president Fernando Lugo and his allies and, particularly, for lifelong leftist activist Ricardo Canese, an engineer who may have even authored the very phrase "hydroelectric sovereignty," this term signaled an understanding of sovereignty as intimately related to natural resources and, specifically, the ability of a polity to determine the destination and use of those natural resources. Canese's role in the Lugo campaign and government is emblematic of how crucial engineering expertise is to politics in Paraguay. Whereas the opening vignette revealed an energy sovereignty tied to nonhuman atmospheric processes, hydroelectric sovereignty linked natural resources to international relations and international law. International and treaty law have often been creatively used to limit, rather than strengthen, the access of local communities to their natural resources; this history extends to the very origins of the sovereignty doctrine and of treaty law at the beginning of the New World conquest, when the puzzle to be solved by Spanish jurists was how to legally justify dispossessing Native Americans of their right to self-rule and their natural resources.[20] Natural resources, and not just human populations, are central to the sovereignty (or lack thereof) of colonized, subaltern, or marginalized communities. Though Canese and Lugo and their allies did not mention this deep history, it was the backdrop for their struggles as they asserted the equality of Paraguay to its much larger neighbor and as their complaints targeted the Itaipú Treaty and, by extension, the larger international legal system.

Coming to the Light: An Economic History of Hydropower

Just as the dam was a part of the nonhydroelectric histories of Brazil and Paraguay, it also emblemized the greater electrical context of the region. Unlike the fossil fuel arrangement in the United States (where subsoil resources are privately owned) and in South America (with fossil fuel concessions granted by the state for the purpose of extraction and sale abroad), hydroelectricity was developed for local consumption and is now overseen by state-involved companies with the aim of bringing wealth and development, although it began as a nineteenth-century foreign investment strategy. The point bears repetition: of the myriad energy sources in South America, hydropower looks inward not outward and is predominantly controlled by the local state (or formerly statal companies) not transnational energy corporations. Mirroring the ways road construction opens up new terrain for development and titling, electrification in South America initially followed the established paths of the telegraph and then the railroad and tramway. Once tracks crisscrossed the hemisphere, foreign banks sought other investment opportunities in the 1880s and 1890s and found them in the "Light." Holding companies and local concerns, variously named São Paulo Tramway, Light and Power (1899, Brazil) or Maracaibo Electric Light Company (1888, Venezuela) or Bogotá Electric Light Company (1889, Colombia), whether or not they were owned or begun by Anglophones, formed throughout the continent. Many of them first used steam but then quickly built hydroelectric plants to generate electricity, even before the turn of the century.

The initial continental choice for hydroelectricity merits its own rigorous study but, for our purposes here, has to be summarized as a combination of expertise, hydrocarbon deposit availability/quality, European wars, and then path dependency.[21] São Paulo Tramway, Light and Power was begun in 1899 by Canadian investors and engineers who had just seen the operationalizing of the world's first alternating current hydroelectric plant in Niagara Falls (1895). In 1912, it was subsumed under a new company, Brazilian Traction, Light and Power Company. Known simply as "Light" in Brazil, the company brought massive electrification to the country in the first half of the twentieth century, before energy production and then distribution were placed in state hands under Centrais Elétricas Brasileiras (Eletrobras). During its early years, Brazilian Traction continually built hydroelectric plants to power its expanding tramway network. Growing local familiarity with hydropower, limited coal deposits, and then wartime scarcity combined to make hydro the viable option for energy generation in Brazil.

Paraguay followed late in the footsteps of its neighbors. In 1910, the Paraguayan congress granted a concession to Italo-Argentine Juan Carosio, who then went on to construct Asunción's tramway and first electrical lines. (Electricity-powered tramways had reached Buenos Aires by 1897.) Asunción Tramway Light and Power Company Limited was acquired by the Argentine transnational firm CALT (Compañía Americana de Luz y Tracción) in the 1919, which laid into the streets of Asunción metal tracks that are still visible today underneath the worn asphalt in parts of the colonial city center.[22] For decades, CALT provided electricity and infrastructure to Paraguay until it was nationalized and superseded by the newly created Paraguayan state company Administración Nacional de Electricidad (ANDE) in 1948. ANDE deepened the reach of electrical lines into the rural hinterland of Paraguay, but the tramway fell into unplanned obsolescence. In the capital of Asunción, the skeletal remains of rail lines in streets now choked by diesel-spewing buses hearken to a distant halcyon past. By the time the Itaipú project commenced in the late 1960s, local governments in Argentina, Brazil, and Paraguay had taken over the responsibility to provide and develop energy resources for local consumption. In fact, Itaipú heralded an important transition from foreign transnational dominance in electricity provision to national state leadership.

Although environmental anxieties around sustainability were not central issues in the mid-twentieth-century spate of nationalizations, concerns about national sovereignty and the responsibility of the state to paternalistically guide the growth of the nation were. Since its inception, Itaipú has remained in the center of an understanding that "the state," endowed with responsibility to and by "the nation," should guide the use of the nation's patrimony to secure the prosperity of the nation not just in the short term but for future generations. Through building Itaipú, the government of Brazil in the form of Eletrobras took a decisive step to provision national industry with secure, reliable electricity. Even the military government of Alfredo Stroessner, under whose watch the dam was built, used the language of patrimony and national development as Paraguayan sovereignty to describe the Itaipú project. As Jessica Cattelino found in her work on Seminole gaming as a defense and expression of sovereignty and as I have written about regarding Paraguayan leftist activism for hydroelectric sovereignty, this is an exercise of sovereignty whose benefit should go to the nation and not to outsiders.[23] But the anxiety to preserve Paraguay's sovereignty evinced by President Lugo, his government, and the Paraguayans who voted him into office had deeper historical roots than the coming of the Light.

Many early hydroelectric plants provided the power for export-oriented mineral extraction, much in the same way that railways developed alongside coal. By the mid-twentieth century, dams like Itaipú were built to power national development. Because of its lengthy history, Itaipú endured changes in reigning economic ideologies in the region. In fact, it sheds light on the rationales for divergent economic policies. The dam began during a period of Import Substitution Industrialization (ISI) in South America (1950–80), a response to the failure of liberal promises of export-led growth bringing economic development.[24] Under ISI, the state fostered economic growth by replacing imports with domestic production through protective tariffs and other industry-sponsoring measures including the outright development of industry by the state itself.

Paraguay did not embark on ISI for economic growth. Instead, the policy of the government has been to encourage agribusiness export-led growth: yerba mate in the nineteenth century and early twentieth century, then cotton, sugar, and now soy via generous land tenure policies benefitting large (often foreign) landholders.[25] At the time of Itaipú construction (and arguably even today), Paraguay was largely a bimodal agrarian society with a handful of elites and a large population of peasants and agricultural workers, attended to by the military and a caste of lawyerly technocrats. Much of the middle class that exists today traces its origins to the new disruptive opportunities (engineering and other professions) offered by Itaipú.[26] These sectors form the basis of what Kregg Hetherington termed the "New Democrat" subjecthood in Paraguay—urban, university-educated, upper-middle-class professionals who grew up during the Stroessner regime but longed for a transition into a liberal democracy and a move away from traditional clientelistic politics.[27] Brazil, on the other hand, was one of Latin America's ISI success stories, having weaned its dependence on North Atlantic agro-import markets for economic prosperity. The period from the late 1960s to the early 1970s, known as the "Brazilian miracle," saw economic growth rates of 10 percent per year, followed by a mild cooling off period between 1974 and 1980, when the economy nevertheless grew at an annual rate of 6 percent.[28] Despite the economic stagnation that followed in the 1980s, the broadening of the Brazilian middle class and the vitality of the early twenty-first century date to this period.

Itaipú was intended to play a key role in Brazil's industrialization, powering it on a scale hitherto unseen.[29] The dam then weathered the decade of economic contraction in the 1980s and the neoliberal experiment that followed.[30] In an era where new economic orthodoxies gripped the continent and resulted in the privatization of state-owned companies, including energy concerns, and where the state receded as private providers stepped in to meet the needs of citizens-cum-consumers, Itaipú was a notable exception. It was never privatized. That Itaipú remained state owned in an era of neoliberal political economics reveals complexities within the global capitalist system and points to how binational integration reinforces the control of state actors because private corporations have hitherto been unable (or unwilling) to absorb the infrastructural investment and risk necessary to construct megaprojects. And, in fact, I would argue that these particularities are not aberrations but, instead, how the system itself works through creative flexibility and dynamism when it comes to region-scaled natural resources. Per Giorgio Agamben's invocation of Carl Schmitt on the exception as key to expression of sovereignty, much scholarly attention has been placed (rightly) on the connection between state-sanctioned violence and states or spaces of exception in modern rule, from the Nazi death camps to Guantanamo in the War on Terror. But Itaipú signals the importance of creating a financial-legal, not necropolitical, exception. Aihwa Ong, for example, has highlighted how offshoring within neoliberalism draws on spaces of exception.[31] Itaipú's ongoing status as a state-owned financial exception also demonstrates how a patchwork of financial-legal exceptions, some neoliberalized, others not, buttresses the global capitalist market.

Widespread dissatisfaction with the outcomes of neoliberal policies contributed in large part to the tide of left-leaning governments, like Fernando Lugo's, that swept across the continent. From Argentina's Nestor Kirchner to Venezuela's Hugo Chávez, leaders came to power in the first decade of the twenty-first century on promises to redistribute rents to populist masses and not just to privileged elites. Especially in Bolivia, Ecuador, Peru, and Venezuela, a Bolivarian social[ist] revolution financed through fossil fuel extraction was seen as the cure to centuries of social inequality. Brazil and Paraguay also participated in the leftward turn, although both were interrupted, interestingly, via parliamentary impeachments. Thus, the defense of national sovereignty through environmental (and economic) sustainability under the benevolent stewardship of the national state was understood as a corrective to the extraction-based colonial insertion of South America into the global economy, which instead prioritized the short-term gain of foreign interests resulting in the permanent destruction of local resources through mining, harvesting timber, and drilling. This anti-imperialist conceptualization of energy and underdevelopment differs from the resource curse narratives that attributed the unexpected poverty of resource-rich regions to the quick wealth of extraction rather than an economy-lifting wider industrialization. Nevertheless, political ecologists from Latin America have criticized the left turn's natural resource politics as neocolonialist "progressive extractivism," incorporating many of the same practices and values as previous governments, merely under a new guise.[32]

Hydroelectricity in the Leftward Turn

The central role of hydroelectric-extraction-led social development in Paraguay's left turn shows a more complicated picture of how groups across Latin America responded to neoliberalism by rethinking the political ecology of the nation's natural resources. I first went to Paraguay in 2007 to study border dynamics, globalization, and the government at the infamous Triple Frontier, where Argentina, Brazil, and Paraguay meet. But, in spite of my enthusiasm for what seemed (to me) like a lively topic on smuggling and the Western Hemispheric outpost of the War on Terror, people insisted on talking about water and repeatedly redirected conversations to energy and dams. Most Paraguayans with whom I spoke appeared uninterested in the rumors of Hezbollah activities or Al Qaeda training camps in the border city of Ciudad del Este or even the local resistance narrative that saw nefarious intentions to seize the water of the "Iraq of the future" behind these rumors.[33] They were instead caught up in the thrilling presidential campaign of a party-less leftist bishop and in the inflamed debates on what should be done about Paraguayan hydroelectric sovereignty in Itaipú Binational Dam.

Election day April 20, 2008, dawned a beautiful, cloudless summer day in Paraguay—warm, but not sweltering, the kind of Sunday spent grilling with the extended family. Tensions and fears ran high as the most contested presidential election in Paraguay's history began, yet years of a brutal military government that had been ousted in 1989 not by a popular demand for democracy but by another coup lent a peculiar character to the public expression of anxiety in Paraguay: normalcy. "Who do you think will win the election tomorrow?" I had asked multiple people the day before. They gave the same answer: "I don't know. The numbers are too close."

The numbers were not close. Though former bishop Fernando Lugo was ahead in the polls, after more than six decades in power the Colorado Party controlled public services, including the electoral commission and the buses to take people to voting stations. Stories of fraud streamed in steadily. But if calm indifference on the streets of Asunción masked anxious engagement, *ABC Color*, the country's newspaper of record, was less timid. As the day began, the paper captured the mood of the country with a full-color political cartoon on the front cover of Lugo finishing a race first, joyful. Emblazoned above the fold and elaborated in the first eight pages, bold prophetic letters announced the theme for the paper: "Brazil exploits Paraguay in Itaipú, this should be the top priority of the next government."[34]

Leftist former bishop Fernando Lugo bested the Colorado Party in April 2008 by capturing and channeling discontent among Paraguay's citizenry over unfulfilled promises and the status quo, linking all these complaints to Paraguay's binational hydroelectric dam, Itaipú. Despite triumphalist imagery, in Paraguay Itaipú stood for two failings: a source of illegal enrichment for well-placed Paraguayan elites; an unequal relationship of exploitation by Brazil in unevenly shared control of a supposedly equal resource. The progressive campaign pledge to recover hydroelectric sovereignty helped catapult the leftist bishop and his allies to power. Electoral success and the progressive rearticulation of hydropolitics forced other parts of the political spectrum to reconfigure their own approach to the issue and to popular discontent. Paraguayan citizens did not merely vote; they continued to put pressure on the government through demonstrations, public questioning, and the engagement of the media. One of the benefits and challenges of my research topic was that everyone with whom I spoke voiced an opinion on Itaipú in Paraguay. It was a cipher people used as a way to talk about everything else in the country: education, land tenure, economic growth.

New promises of energy rent redistribution mimicked the petropopulisms of Bolivia and Venezuela, but with repeated emphases on Itaipú as a "clean," non-carbon-emitting source of power. Thus, the leftward turn spreading across Latin America reached the landlocked country at the heart of the continent. Notably, the principal progressive social concerns connected to Itaipú during Fernando Lugo's campaign and presidency were not the people displaced by the flooding of the reservoir (indigenous communities and precarious campesinos who were left unpaid and dislocated) nor the construction workers who claimed to never have received their salaries. Instead, the hope was to strike a balance between market-based exportation of electricity, which would then finance social transformation. Left-of-center governments in the region looked favorably on existing and potential hydroelectric projects in spite of critiques of dams as ecologically and socially destructive by campesino, indigenous, and environmental groups. While the headline-grabbing governments of Venezuela and Bolivia had much to do with the management of fossil fuel natural resources, hydropolitics allows us to see the formation of new political constituencies that advocated state-directed energy-based economic growth and social development. In fact, in the first decade and a half of the twenty-first century in Paraguay, Itaipú hydroelectric resources were at the center of a contest between two groups of political actors: market-friendly or, at least, tolerant socialists like Fernando Lugo himself and his close ring of advisers; and advocates of state-checked liberal economic policies, like Carlos Mateo Balmelli and other members of Lugo's short-lived governing coalition. Notably, both groups favored an interventionist state.

Water as energy, with constraints both technical and geographical, effects and affects state-making processes. Harnessing hydroelectric resources has left a particular imprint on political and economic structures within countries that depend greatly on dams for energy, what I have referred to as hydroelectric statecraft.[35] Because the debt-financed costs of constructing dams run so high and because the lead time until energy production is so long, often stretching into decades, few organizations other than the territorialized nation-state possess the financial and institutional wherewithal to commence and complete projects under these conditions. The materiality and management of water have resulted in distinct hydrostate-making forms that parallel (though differ from) petrostate or carbon democracy forms.[36] When industrialization and economic development fueled by hydroelectricity are contingent on and consequently subordinated to state actors and state institutions, they produce "hydrostate effects."[37] The place of state power in hydropower is only augmented by the kind of transportation required for hydroelectricity. Unlike fossil fuels, which are stored and shipped, hydropower is generated and instantly consumed via high-tension wires, meaning that the territorial base of hydropower is all the more constrained. One key hydrostate effect is the intersection of risk, temporality, and uncertainty. All hydroelectric dams have limited operative life spans (Itaipú was initially rated for 100–150 years), but planners may overestimate functionality because there are too many unknowns about the future (climate change, geological instability, economic crisis, war). Itaipú was built to provide electricity but grew to assume a management role coordinating electrical systems in both Brazil and Paraguay (what engineers call "dispatch services") and now, especially because of climate change, managing water.

Perhaps because politically attuned Latin American anthropology has elucidated how the region is inserted unequally in the global economy, much of the early groundbreaking work on the state in Latin American focused on the impact of fossil fuels and mining in Latin America.[38] Yet one conspicuous feature of the hydrocarbon resources in the region is that they are a strategy of extraversion.[39] Hydroelectricity is extracted and consumed within the continent and thus has regional hydrostate impacts that lead to ecocitizenship relations that disrupt nation-state boundaries. Decades ago, Laura Nader, as part of the growing anthropology of energy, encouraged ethnographers to "study up" and to study energy. Her work on energy sector elites revealed how scientific expertise and careerism translated into energy options for the US public.[40] Building on these early interventions, more recent work on nuclear power has shown how the size and security implications of Los Alamos transformed the United States into an atomic nation.[41] On the renewable energy front, new scholarship has studied how local aesthetic obstacles to wind result in corporation-benefitting offshoring, and although microhydro may have more potential for community empowerment, it may reproduce racial and caste hierarchies.[42] But Itaipú Dam is neither so nimble nor so local.

The political, social, and ecological implications of Itaipú are all the more pressing in the wake of a flurry of new hydroelectric dam construction across the globe, a move rhetorically celebrated as a turn to clean energy, distinguished from carbon-emitting fossil fuels.[43] A twenty-first-century anxiety over menacing water wars raises the stakes about who has the right to determine and deny water usage.[44] For years, literature on dam effects has documented dispossessed, displaced peoples and environmental devastation, unintended consequences that resulted from a failure to fully consider ecosocial impacts.[45] And yet, scholarship suggests that the latest hydroelectric feasibility studies worldwide still neglect public health, labor economic, and environmental science data that document how hydroelectricity is not a simple panacea for poverty. Anneliese Petersen, for example, argues the planned Inga III Dam in Congo may result in a flare-up of schistosomiasis, a "disease of poverty" unmentioned in the World Bank's favorable feasibility study of the proposed dam.[46]

Instead of expressing environmentalist politics, the political economy of hydroelectricity in Africa, Asia, and South America has powered the growth of national and transnational industry and the agendas of local and global elites. And while dams are often legitimated by a utilitarian calculus of "greatest number benefited," the fact remains that Three Gorges (China), the Mekong River basin dams (southeast Asia), and Yacyretá (Argentina-Paraguay) to date have underdelivered on energy production promises. The business of power has made powerful people even more powerful. With exploitative effects beyond land expropriations or unpaid labor (though these are part of the history of Itaipú), hydroelectric dams can be used by government elites to mete out state violence and human rights violations.[47]

One question is how to even account for the political, social, and environmental costs of hydroelectricity. Here, the concept of "externalities" from business economics proves useful. An externality or "spillover" effect is a benefit or cost incurred by a third party that did not choose to incur that benefit or cost. Economic anthropology has shown that the lines between externality and cost are culturally constructed and not set according to consistent rule.[48] By definition, externalities are not included in the cost of doing business and, instead, are offloaded elsewhere. Although per the Itaipú Treaty, the displaced populations, flooded land, and lost productivity were to be compensated by the dam, the history of Itaipú reveals some losses were compensated as promised and others were not. Itaipú Dam's externalities have as much to do with conjunctural political expedience and, as historian Jacob Blanc demonstrates, racialized hierarchies as they do blueprinted plans.[49]

Ecological crisis and strained natural resources have demanded a new assessment of environmental externalities, and pleas for sustainability often include a call to "internalize" an externality. Renewable energy has different internalized energy externalities than fossil fuels, an undercurrent that runs throughout this entire book. Therefore, Itaipú not only helps us understand renewable energy; it also serves as a foil, reflecting on the political economy and ethical assumptions about fossil fuels. Rather than lionizing the Brazilian-Paraguayan dam as a miracle of development or decrying it as a destroyer of natural and social environments, this book is a more targeted discussion of why people have chosen hydropower and the outcomes thereof because, whatever the drawbacks, hydro remains a significant source in South America. Even progressive political activists in Paraguay, who were critical of the political-economic outcomes of neoliberal energy export policies, nevertheless affirmed the positive contribution of renewable energy extractivism as part of left-turn regional politics.

All energy sources have some kind of liability: fossil fuels are finite; anthropogenic climate change is the Achilles' heel of hydropower. Indeed, part of Itaipú's importance arises from how the dam figures into managing risks from climate crises, which are often configured as a source of state insecurity and social vulnerability.[50] Since completion of construction in 1992, Itaipú Dam has repeatedly set the world record in energy generated. Though Three Gorges Dam in China now has a greater installed capacity of 22,500 megawatts, it produced less electricity than the Brazilian-Paraguayan dam two-thirds its size until the drought of 2014 sent Itaipú's production plummeting for one year only before it bounced back. Record-breaking months of poor rainfall drained reservoirs in southeast Brazil, and water rationing drove urban residents of São Paulo to drill wells into their basements. The already festering political crisis that culminated in the 2016 impeachment of Brazilian president Dilma Rousseff was exacerbated as industry competed with agribusiness and with ordinary citizens for hydro resources. In normal years, 80 percent of Brazil's electricity comes from its extensive network of renewable energy sources (hydro, wind, solar).[51] But the 2014–15 drought forced the country to rely on its stopgap thermal plants more than usual. And so, for 2014 (the same year that Itaipú posted below-average production numbers), only 70 percent of Brazil's electricity came from renewables.[52] By 2016, Itaipú once again broke world records for energy generation.

These insecurities and the discovery of offshore petroleum reserves have led some to foresee a diminished role for hydropower in Brazil's future, with the expectation, perhaps, that the rest of the region will follow suit.[53] Megadam construction experienced a new heyday during the left-of-center Brazilian presidencies of Lula da Silva and Dilma Rousseff. At its height, some 28,602 megawatts of additional hydroelectric capacity was in the works—two of the dams, Belo Monte (11,223 megawatts, to be completed in 2019) and São Luiz dos Tapajós (8,381 megawatts, unconstructed), were planned on the scale of energy generation that Brazil purchased from Paraguay's half of Itaipú or more. But in January 2018, the new conservative government in Brazil that replaced Rousseff announced the halt of megadam construction, especially those planned for the Amazon, in favor of smaller hydro, solar, and eolic sourcing.[54] Indigenous and environmental groups have celebrated. Indeed, Brazilian president Michel Temer attributed his government's decision to suspend the projects as the result of the will of the people.

However, it may have as much to do with the massive corruption scandal that engulfed the Brazilian government, tied to bribes paid out by large construction companies, including Odebrecht, which has put pressure on government-initiated infrastructure projects. The move also coincided with Temer's plan to privatize the majority of Eletrobras, further evidence that megadam construction projects are tightly associated with an interventionist state but not with private enterprise (smaller hydro, wind, and solar all have both shorter construction time spans and lower costs). If that is the case, it is imaginable that, once the increased scrutiny of construction projects passes out of fashion or with the election of a left-of-center government in Brazil, these scuttled projects may find new life.

Researching Hydropolitics from the Margins: Anthropology in Paraguay

Paraguay is a source for creative political experimentation around energy sovereignties, from reliance on nonhuman atmospheric processes to a recognition of how subaltern sovereignties are limited by international law. The case of Itaipú Dam in Paraguay speaks to the engineering of modern South America because of the prominence of hydro-sourced energy in the region, because of the ways extractivism—from mining in the colonial era to agribusiness in the present—has structured economics and politics in the (post-)

Iberian world, and because of the weight of the international legal structures built around Itaipú. It may seem odd that the first scholarly monograph in English to analyze the political economy and political ecology of South America's largest dam comes from an anthropologist. But perhaps this reflects the way power works in Itaipú and Paraguay; it most certainly points to the unique aperture allowed by the progressive Lugo government. Because it lay outside the jurisdiction of congress and comptrollers, the more typical bureaucratic avenues to the governing heights were closed off. Rather than submitting a research proposal request to a government agency in Paraguay, I encountered a state apparatus that was highly personalistic. And so, I was entirely dependent on some persistence, a great deal of good fortune, and an even greater deal of generosity on the part of people I met in Paraguay who invited me into one of the most controversial issues of the day. It was through Marilín Rehnfeldt, anthropologist at the Catholic University of Asunción, that I met the leadership of Tekojoja, the leftist social movement that launched Fernando Lugo's successful presidential campaign. Because of her, I spent part of election day 2008 at Tekojoja headquarters, visited voting stations (where I unwittingly photographed voting fraud as a uniformed police officer cast his ballot), and even got behind-the-scenes press access to Lugo as he declared victory and ended one-party rule.

Key leaders in Tekojoja—activists who had been persecuted, surveilled, and exiled in their youth at the hands of Alfredo Stroessner's military government—had spent decades protesting the development and administration of Itaipú. They put the dam squarely on the campaign agenda of the liberation theology bishop. Ricardo Canese, an engineer who was one of Lugo's closest collaborators, led the hallmark progressive movement to recover Paraguay's hydroelectric sovereignty by demanding a renegotiation of the wealth- and power-sharing relationship with Brazil in Itaipú. With Lugo's victory, Canese was at last authorized to tackle the issue to which he had dedicated four decades of his adult life. He and other leftist activists were appointed to a special Hydroelectric Commission within the Ministry of Foreign Relations to spearhead the contentious negotiations with their Brazilian counterparts. As members of the market-friendly Left, their aim was to connect with other popular sectors in Brazil. For part of my fieldwork on energy politics in the Lugo transition, I attended public presentations of the commission and conducted private interviews with its members from 2008 until I returned to the United States in 2010. Ricardo Canese, his team, and Fernando Lugo were instrumental in shifting the discourse on renewable energy in Paraguay toward a Bolivarian framing, a combination of liberal and leftist economic practices where market-based energy rents are used to finance social development and the implementation of socialism.

The circumstances by which I gained access to the Executive Directorate of Itaipú were equally fortuitous. I visited Itaipú headquarters in Asunción weekly for fourteen months, conducting extended interviews with engineers of both of Paraguay's major binational dams and multiday visits to Itaipú Dam. I was able to interview the Paraguayan executive director dozens of times, to speak to all the Paraguayan directors and many of the Brazilian directors, and to interview and observe the staff and the financial, legal, and engineering experts that ran the Executive Directorate. I gathered hundreds of hours of conversations and documents, both published and unpublished, about the dam. I even helped give presentations to foreign visitors and was interviewed by the Paraguayan press myself. And all this began with a protest on a November morning seven months after Fernando Lugo's electoral victory where I was among a group of two hundred who descended on the Asunción office of Carlos Mateo Balmelli, executive director of Itaipú.

Lugo's win came as the result of a fractious coalition that included the Liberal Party, the traditional rival to the ruling Colorado Party, so that when the time came to divide up the cabinet-level leadership positions of the government, Lugo appointed Liberal politician Balmelli as executive director of Itaipú. Balmelli and the team of energy sector technocrats he headed advocated public-private partnerships and market-led growth under the guidance of the state as well as more traditional welfare interventions in order to bring social development. As someone familiar with the dissertation process (he had a PhD), Balmelli invited me to regularly visit the headquarters in Asunción to learn more about the changes in the dam, saddling me with a stack of books that he thought would be necessary background reading, and even asking the communications department to set aside a desk for me. Whereas the success of the Left lay in popular discourse and in changed objectives at the highest level of government, the liberal democratic vision gained traction on an international plane and among business sectors in the region (chapters 3 and 5).

The two Paraguayan energy teams had differing outcomes in their negotiating strategies in part because of how and where they applied pressure and because of how they framed their goals. The Ministry of Foreign Relations leftists connected with popular groups in the Southern Cone and served as the public face of new energy politics in Paraguay. Public pressure and a wide realignment of political values propelled by Lugo's progressive campaign altered hydropolitics. Because of these changed expectations, energy sector experts were able to draft the negotiated resolution with Brazilian political and economic leaders. Itaipú technocrats developed legal and energy proposals that became the Joint Declaration (2009) signed by Brazilian president Lula da Silva and Paraguayan president Fernando Lugo, codifying new dimensions in binational energy governance. And yet, adoption of any solution depended on the leftist solidarity between the two presidents. Soon after, partisan machinations within Paraguay impeached Lugo in what some called a constitutional "coup," paving the way for the return of the Colorado Party elites in the executive.

With welcome from two very different (and sometimes even antagonistic) quarters of the Paraguayan energy sector—the progressive activists in the Ministry of Foreign Relations and the more traditional technocratic experts within Itaipú and ANDE—I was able to witness energy policy unfold nationally and internationally from 2007 to 2010. I also conducted research with Brazilian government officials and energy experts and follow-up visits in 2013, 2016, and 2017. As with most ethnographic research projects, I was led along a different path than the one I had originally laid out and found myself studying and writing about technocratic expertise and institutional culture as much as about sovereignty and political economy. But even as the ability to direct the material and symbolic resources of Itaipú offered an unheralded ability to redirect the development of the state within Paraguay, much more was at stake within the dam. Itaipú reflects a regional shift in energy governance. And this was why the new negotiations in Itaipú became an inflammatory topic in the Brazilian congress and in financial newspapers in Brasilia, New York, London, and Madrid, and why politicians across the Iberian world kept tabs on the proceedings.

In many ways Itaipú was a "total social fact" in Paraguay. Early anthropologist Marcel Mauss used the phrase to describe phenomena that are simultaneously legal, economic, religious, aesthetic, and so on.[55] For instance, in coal country in West Virginia, coal dominates how locals (those who work in the industry as well as those who do not) talk about and do human community. Similarly, in Paraguay, Itaipú—as energy, as a source of wealth, as international political heft—was, as Mauss wrote, "more than a set of themes, more than institutional elements, more than institutions, more even than systems of institution divisible into legal, economic, religious and other parts."[56] This is why trained engineers became a necessity in politics, why everyone I spoke with (nonexperts included) had an opinion on the subject, and why the study of energy infrastructure in Paraguay allows us to analyze how power works in Paraguay.

For each of the chapters that follow—in Itaipú's engineering (chapter 1) and financial (chapters 2 and 4) dimensions, in its role in extending Latin America's leftward turn in Paraguay (chapters 3 and 5), and in scripting a South American region via energy integration (chapter 6)—the dam offers the opportunity to see how decisions are made that establish a hierarchy of values. The effects of these choices result in a political economy with hydroelectric contours that reveals financial priorities and state-making prowess predicated on and simultaneously constructing new political-communal relationships to nature. Moreover, if the resurgence of megahydro was a feature of left-turn politics, then this book reveals dynamics about hydropowered industrialization that cohere with left-turn politics; that is, it uncovers why left-turn politics should be hydropolitics. Subterranean but powerful anxieties around scarcity, destruction, and national development motivate the elevation of sovereignty and security as critical components of hydroelectricity. The solutions to these concerns, we will see, amount to a seismic shift in how energy is done.

PART I
Circulations

1

Current

Introduction

In June 1973, just as they were in the thick of debate over the proposed Itaipú Treaty, members of Paraguay's Chamber of Deputies received a lengthy open letter circulated by a group of engineering students. Appealing to the deputies' love for and fidelity to the nation, the students wrote: "We hope that this sincere and impartial analysis from young Paraguayans, defenders of the nation, may serve you as a true representative of the PEOPLE . . . [to understand] the true value of the Itaipú Treaty and its consequences for the development and the integrity of the country, the lack of clarity of which, with its dual interpretation and real disadvantages may provoke for Paraguay."[1]

What followed were ten pages of precise details about turbines, the frequency of electricity generated, nineteenth-century border agreements, international law, currency demarcations, and institutional hierarchy as decision-making procedure—the typed-up results of a two-day conference hosted by the Center for Engineering Students (CEI) at the National University's Engineering Faculty on the newly circulated Itaipú Treaty. For a discussion held at the country's premiere engineering department, a surprising amount of the debate centered less on mechanical information and much more on the interpretation of legal precedent. Like the engineering dilemma from the introductory chapter as to whether to allow upriver dams to provide water for increased Itaipú production, the Center for Engineering

Students understood Itaipú electricity and the mechanics required for generation as simultaneously technical and political questions. The conclusion of the letter was clear: the engineering and administrative stipulations as written in the treaty might "impinge on our sovereignty."[2]

I found this document, with slightly yellowed paper and smudged carbon-copied ink, not in any engineering archive or congressional archive, but rather at the Museum and Center for the Documentation and Archive for the Defense of Human Rights. Better known locally as the "Archive of Terror," the Asunción archive is a five-ton collection of interrogations, arrest records, audio recordings, confiscated personal items, and medical reports detailing hunger strikes, torture, and forced miscarriages gathered by the Ministry of Interior's police and investigations units during the military government of Alfredo Stroessner. The now public archive was an excellent source of documents on Itaipú Dam. And, clearly, the open letter from the engineering students was deemed important enough to national security that it was filed alongside surveillance reports of political dissidents. Hydroelectric dam construction worldwide is frequently controversial; in the Itaipú case, trained engineers got involved in the strife.

The engineering student debate and letter show that the Executive Directorate's decision to wait for forecasted rain rather than recur to an engineered solution forty years later was only one example of a myriad of sociopolitical problems associated with energy sourcing. From the very beginning of construction, disputed engineering decisions were seen as jeopardizing the sovereignty of Brazil or Paraguay (or even other neighboring countries). I was once told by a senior engineer tasked with building the original cement factory for the dam in the 1970s that "power *is* politics." Though I had not appreciated all that this meant at the time, I could tell by the engineer's demeanor that he was saying more than a pithy, but tautological, sound bite in that sentence. To an engineer, "power" has a specific definition. Power, measured in watts, describes a quality of electricity. More specifically, power is the rate at which an amount of energy flows per unit of time.[3] And so, in saying that power was politics, the senior engineer asserted that the megawatts produced by Itaipú were units not just of energy but also of something else, something political. Like the engineer, in this chapter I argue through the engineering necessary to generate current that Itaipú power is politics.

As Itaipú became steadily more important to Paraguayan politics and economics during planning and construction, engineering and engineers themselves became indispensable to the doing of politics. Here I turn to engineering decisions during Itaipú construction that were construed by engineers as having to do with sovereignty in order to learn to read the current, to show how Itaipú energy works by showing that any watt anywhere is a mechanical and sociopolitical thing. Concomitantly, engineers and engineering became an indispensable part of governance. In the hands of energy engineers, the tools of scientific inscription not only made hydrology legible; they transformed weather into politics and, thus, hydropolitics.[4] Because of the ongoing national and international context of Itaipú, technical details that otherwise might be deemed boring or inscrutable caught the attention of Paraguay's public and governing elites. The engineering students who signed the open letter went on to run engineering firms, not pursue government careers. Nevertheless, they became interesting to the regime as applied science became controversial and calculations themselves threatened the integrity of the political system.

The energy managers responsible for planning and constructing Itaipú were given two tasks: they had to make megawatt hour units of energy and they had to make them national. They did both through the construction of retaining walls, the size of turbines, and the technical specifications of transformers, that is, through engineering interventions. Not only did Itaipú technicians have to balance complicated equations; they had to be able to interpret the constraints from physics and the design of Itaipú according to political values. The first section of the chapter offers a technical explanation of just what a watt is and how Itaipú generates electricity in order to ground the basic argument that the materiality of the form of energy (in this case, water) is intimately tied up with the political forms connected to that energy source.[5] I briefly describe how Itaipú generates electricity, discussing the peculiarities of renewable energy vis-à-vis hydrocarbons, in order to understand the considerations and constraints weighed throughout the book by expert energy managers, by Brazilian and Paraguayan politicians, and by civil society. Perhaps because most people are more familiar with how wood fire releases heat for warmth or cooking, forms of energy that come from ignition—for example, burning fossil fuels—seem easier to conceptualize. A hydroelectric dam is a less intuitive source of energy; water is, after all, often imagined as the opposite of fire. And, outside of dramatic lightning strikes, electricity is unseen to the human eye. Electrical lines are lifeless, inert, until something invisible happens to them, animating them to power machines and rendering them lethal to the touch. Itaipú energy requires some reverse engineering.

But more is afoot in Itaipú. Co-owned by one of the world's largest countries and one of Latin America's smallest, Itaipú Dam has internalized the asymmetries between Brazil and Paraguay in spite of attempts to balance equality. Itaipú is not just any hydroelectric dam, and the engineers and administrators who run it are not ordinary public utility employees. Because the gargantuan dam is binational, energy production is a matter of constructing sovereignty and takes on diplomatic overtones. The engineers had to find ways to give the electricity—that is, the flow of electric charge generated by the same water of the same river at the same dam—two discrete national identities as a way to defend the sovereignty of the dam's two owners. Electric charge is a nigh immaterial phenomenon; electrons are elementary particles, with no known components or substructures, and are effectively fungible. Yet there are "Brazilian" megawatts and "Paraguayan" megawatts. Through material and symbolic entailments, Itaipú engineers have constructed energy sovereignties by bestowing sociopolitical identity on electricity and thus fashioned a territoriality that emanates from the circulation of charge.

Because of the physics of hydroelectricity and because of the specific history behind the dam, the nationalistic stakes of Itaipú are ever present. That is to say, even though electrons have no known parts, the cultural geography of water and the national histories of Paraguay and Brazil are core components of Itaipú electricity, requiring that technicians and technocrats superintend Itaipú energy decision making as international political processes. Decisions about how to optimize flow or whether to build a new transmission line necessitate international negotiation because they impact national sovereignty and thus accord Paraguay more international leverage than usually warranted for a small, landlocked country overshadowed by powerful neighbors. Itaipú energy management has alternatively functioned as a proxy for and/or competitor to the foreign ministries of both countries, to which it is legally subordinated. And so, in the second portion of the chapter, I focus on two major design decisions—having to do with the location of turbines and the frequency of electricity generation—that reveal how (and why) Itaipú megawatts came to possess nationality and suggest why members of Paraguay's government and energy sector saw machinery as a question of sovereignty. The results established legal-engineering precedent for how to delineate national territory within circulations and potentials and set the stage for the binational negotiations in the Fernando Lugo era.

At the heart of chapter 1 is a technopolitical formation: how state experts and state expertise are reconfigured through the production of energy. By controlling the current in order to resolve political problems, through the construction of the dam itself, engineers in Paraguay attained a notable public standing and political-economic identity. The prominent place of leftist energy engineers like Ricardo Canese in Fernando Lugo's presidential campaign speaks to the general role of engineering in Paraguayan politics. And though Carlos Mateo Balmelli, former president of the Paraguayan senate and a presidential candidate for the Liberal Party, was a career politician, the other energy managers who staffed the Paraguayan Executive Directorate understood their work as inextricably political as well. Max Weber, in his 1919 essay "Politics as a Vocation," explored the functioning of politics outside the mere confines of government employment and instead looked to how many kinds of professions depended on and did politics.[6] He argued, for example, that in the nineteenth century, journalists themselves had become a type of politician, influencing policy and doing the work of politics.

As Itaipú is a binational, government-owned dam, the work of producing energy is no longer just a matter of mechanics. Since engineers must answer political questions, engineering is thus more than a scientific endeavor, rather, in hydroelectricity, becomes a form of politics as a vocation. But to inflect Weber, I am interested in not only what happens to science when it becomes a site of politics, but also the reverse: what happens to politics when you do science in it? For instance, as state security risks become scientific problems, political crises are discussed as questions of differential water sourcing (river or rainfall) and climate change disruption. These conversions—of engineers into politician, of politics into science, of hydrological forecasting into sovereignty, and of circulating charge into national territory—are crucial examples of how Itaipú Dam produces hydrostate effects.

Infrastructures tend to "black box" political relations.[7] Brian Larkin has made a similar argument about opacity as a hallmark trait of infrastructure, which is normally invisible, taken for granted, and even naturalized until something goes wrong.[8] According to this read, crises and breakdowns (such as blackouts or Fukushima's flooded nuclear core) pull back the curtain and make evident those taken-for-granted technical and social elements whose functioning would otherwise remain hidden. Because moments of construction, failure, or ruin disrupt infrastructure's invisibility, they lend themselves to ethnographic and historical analysis.[9] Nevertheless, to limit analysis to extraordinary moments of construction or to system failure allows the occluding quality of infrastructure to cloak political processes. For both the student engineers in the 1970s and the Paraguayan public that supported the push to recover hydroelectric sovereignty in the 2000s, Itaipú political relations were anything but invisible.

Energy resources like Itaipú are politically enmeshed; they may consequently suffer the same flaws of the state structures that administer them. This extends to state violence, not just state sovereignty struggles, when megaprojects are so frequently undertaken by authoritarian governments, as was Itaipú in its construction under the right-wing military government of Paraguay's Stroessner (1954–89) and Brazil's right-wing military government (1964–85). The historical trend that predisposes centralized power (nuclear plants and massive hydro) to centralize power (political) with often antidemocratic results raises the question of just how deeply linked the two are. And if, in its designedness, Itaipú Dam not only reflects but fosters values, then the ethical entanglements of megadams trouble all the more. Whether run by civilian or military governments, reliable access to energy is a matter of national security and defense. Decisions of all kinds served as inflection points where values became embedded. Indeed, there is no politically innocent science. Thus, energy expresses and reinforces the priorities of a human community even as Itaipú energy is haunted by the climate change brought forth by the burning of fossil fuels.

Making a Megawatt Hour: Supply, Demand, and Telenovela Temporalities

The task of the electricity industry is, at its very core, that of balancing watts and wants. Itaipú energy managers must harmonize nature (the amount of water available, the rules of physics) with culture (patterns of human activities, industries, responses to the weather). Electricity works exclusively through circulation—the pull of demand and the correlated push of generation that ensures that, when it is drained, there is always more in that moment to replace what was just used. And a megawatt hour, because it is the product of a push-pull, because supply is created simultaneously with the expression of demand, internalizes both nature and culture, indissolubly blending both. Hydroelectric generation accounts for multiple temporalities: the instantaneous generation of electricity versus the time necessary to fill the reservoir, human work weeks versus rainy seasons, the length of a popular telenovela versus the length of a drought. Teasing apart the way Itaipú produces renewable energy reveals where political problems might come up by undoing the invisibility of both infrastructure and electricity as well as specifically showing why Itaipú energy managers use engineering information to political ends.

The Paraná River begins deep in the national territory of Brazil, where the Paranaíba and the Grande Rivers, two great waterways in their own right, meet. It meanders southwest, watering fields and picking up drainage from smaller rivers until it hits a steep slope carved east of the Mbaracayú mountain range. At this point, the Paraná becomes the border between Paraguay (on the right bank, relative to the direction of water flow) and Brazil (on the left bank). The more than forty hydroelectric dams whose gates might be opened to send water to Itaipú sit on these rivers or their tributaries. Prior to 1982, the Paraná rushed down a precipitous decline over Guairá Falls, which possessed the greatest flow rate of any waterfall on the planet. Over the 105 mile course from Guairá to the confluence with the Iguazú River (the border between Argentina and Brazil), the Paraná drops 393 feet (120 meters). Now, the Paraná is arrested by Itaipú Dam, whose five-mile-long, 656 feet (200 meter) high walls have forced the river to create a huge lake that took only fourteen days to fill when, on October 13, 1982, the gates were finally closed for the first time. Over the day, month, and year, water is stored in the Itaipú reservoir, poised until just the moment it is needed and the intake valves are opened to allow it to enter the generator units. A hundred miles north, the tops of the rocks of the highest Guairá Falls still crest above the lake, testifying to their submerged, resolute existence.

The single most important determinant of the amount of energy produced is the "head," the height the water falls (i.e., the difference between the water level in the reservoir and the water at the river below). Head is related to the hydraulic potential of the river. Excess water is discarded via the spillways to ensure that the reservoir level stays between the minimum necessary for basic operation and the maximum that would overwhelm the system (or breach the dam). Some water makes it through the "run of the river" dam at all times. As soon as the reservoir first reached the 120-meter fill line at the beginning of November 1982, water began to stream down the three-lane spillways west of the turbines.[10] To a lesser extent, the temperature and flow rate of the river play a role in determining the amount of energy a dam can produce. Rainy periods upstream mean that the reservoir will refill more quickly than in drier periods, that is, that more energy can be generated. Consequently, downriver engineers need to be aware of upriver water levels, even those solely within Brazilian territory.[11] And so, to return to the opening of the previous chapter, what the Itaipú Executive Directorate suggested was to make more energy to take advantage of a rise

FIGURE 1.1. Detail of Itaipú penstocks.
Source: Author's photograph.

in demand in the market by opening the intake gates more, even before the rains had come to fill the reservoir.

Itaipú converts kinetic energy into electricity when water pours through vertical sloping penstocks, causing the pinwheel-shaped turbines to spin before the water exits the dam and continues south on the normal course of the Paraná into Argentina and then the Atlantic (see figure 1.1). As Michael Faraday found in 1831, a moving magnetic field induces a current in a conductor; the spinning turns magnets, inducing a current that can be channeled through high voltage power lines and substations. Massive turbines (large enough to seat a full orchestra each, as a construction-era photograph displayed) turn ninety times a minute, leaving the air warm with the murky scent of the lubricating oil that smoothes the rotation of the generating units. The whole floor hums with the vibration of the movement. Each of the dam's twenty turbines has an installed capacity of seven hundred megawatts, enough to power a city of 1.5 million people each.[12]

Because they are rates and capacities, the units to measure and describe the dynamic properties of electricity present an interpretive difficulty; we often turn to comparison with other energy sources or to other physical metaphors. The abstractness of electricity makes it all the more important to understand how it works. Power lines are not like gas tanks. They carry electricity; they do not store it. This is because electricity is a current; it is the circulating flow of electrons and not the electrons themselves. And so, when a light switch is flicked on, at the opposite end of the line hundreds of miles away, electric current has to be ready just in time to travel along the wires to illuminate a light bulb in a dining room. Too little electricity leaves the etiolated bulb dim; too much burns it out. The "pull" of consumption, of demand, must be matched to the "push" of generation, of supply. Both involve constant recalculation, a carefully timed dance between production and consumption for, when users turn on air conditioners, televisions, or machines in an auto-assembly, supply instantly meets demand. Itaipú engineers must anticipate both.

This may seem less intuitive than gasoline in a car or coal for an industrial factory where the fuel sits, waiting to be consumed. In truth, water in a reservoir is akin to fuel in a tank. Whereas fossil fuels are spoken of in volumes, electricity is denominated in potentials: a watt is a rate; a watt hour (or megawatt hour) is an amount.[13] Standard household light bulbs require sixty watts of power to run, but energy experts speak in terms of Itaipú energy megawatts (one million watts) or gigawatts (one billion watts) because watts are cumbersomely small compared to the output of the dam. Watt hours measure the quantity of electricity generated. If a person leaves a sixty-watt bulb turned on for two hours, then 120 watt hours of energy have been consumed. The state of California, the most populous and one of the largest energy consumers in the United States, consumed 281,916,000 megawatt hours of electricity in 2016.[14] In that same year, Itaipú Dam supplied its two owners with 102,335,000 megawatt hours, or enough to meet 36.3 percent of all of California's needs.[15] Customers are billed for the amount of energy consumed: gallons of gasoline in cars, kilowatt hours (or megawatt hours) in utilities.

High-voltage transmission lines carry electricity from the dam to substations throughout Brazil and Paraguay—if watts are a rate and watt hours are a quantity, then volts describe pressure and track the ability of the lines to carry moving current. Because of ordinary efficiency gaps, each turbine does not always generate the maximum possible. Only eighteen of twenty

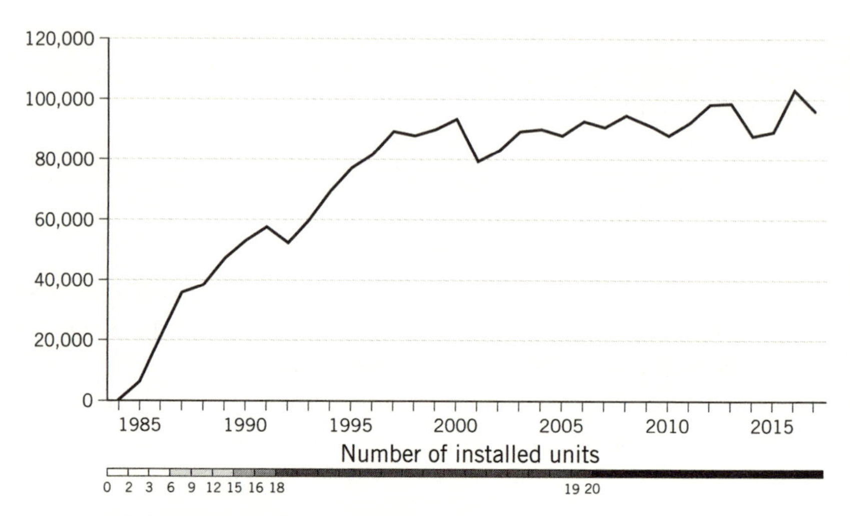

FIGURE 1.2. Itaipú energy production, 1984–2017.
Source: Itaipu Binacional, n.d.

turbines are to be in play at any one moment, meaning that the dam's working capacity is 12,600 megawatts. Although technical staff on tours often explained this as a matter of maintenance, the origin of this arrangement was an international crisis with downriver Argentina in the 1970s. Concerned about the geopolitical rapprochement between Paraguay and Brazil, the government of Argentina threatened to construct a binational dam (with Paraguay) just south of Itaipú and raise the water level at the base of the Brazilian-Paraguayan dam, thus compromising its efficiency by affecting the head. To diffuse the situation, the governments of Brazil and Paraguay agreed to limit the number of turbines in use to eighteen at any single moment, a foretaste of international conflict resolution to come.[16] The 12,600 megawatts, because of the efficiency of the dam, generate record-shattering quantities of electricity. Since the very first turbines began sending current down high-voltage power lines in 1984 to the present, Itaipú has produced 2,514,051,000 megawatt hours of electricity (see figure 1.2). In 2016, Itaipú set another world record for energy generation: the dam produced 103,098,366 megawatt hours of electricity, of which 102,335,000 megawatt hours met 17 percent of Brazil's needs and 75 percent of Paraguay's that year (the remaining portion was used by the dam itself).

But supply is only half of the equation. Although computers now calculate and monitor the careful matching of production-consumption, human engineers oversee and augur changes in the supply of available energy and the demand for it. Technicians in Itaipú are in ready contact with their

Figure 1.3. Itaipú control room. Computers and panels on the top/right control Paraguayan turbines, and those on the bottom/left control Brazilian turbines. Note the line marking the international border on the floor of the room.
Source: Author's photograph.

counterparts in ANDE and Eletrobras regarding changes in consumer demand, desires, and needs. The perfectly symmetrical glass-enclosed control room at Itaipú Dam looks like it could be the set from *Star Trek*. Along the beige walls, blinking lights and needles on dials indicate (in analog) what flat computer screens detail (in digital) on newer workstations: the production details of each turbine in real time (see figure 1.3). A line on the floor marks the actual border between the two countries, with the equipment for the Brazilian turbines on the east side of the line, the Paraguayan equipment on the west side. But the engineers who work at each station may be either Brazilian or Paraguayan, and they rotate through in order to ensure parity. I have never seen a female engineer in the control room, although according to Itaipú publications, there is a small but growing number of female engineers. Like the control room, the staff of the dispatch room, which communicates between ANDE and Eletrobras and the control room, must come from both sides of the dam. Rather than the perfect symmetry of the control room, however, the dispatch room contains much more detailed information about the Paraguayan electrician grid than Brazil's, reflecting a dry technical detail about the location of ANDE's right-bank substation versus the Eletrobras substation. ANDE's substation is fully within the limits of Itaipú, meaning that the binational dam and staff has complete access to information about the Paraguayan market. But the Eletrobras substation lies just outside of Itaipú, sending finalized numbers to the binational entity.

Extreme weather patterns, cool or hot; extended work hours at a factory necessary to fill a large order; the 2018 World Cup final between France and Croatia televised on millions of screens—all require electricity, and all must be foreshadowed. In the UK, energy managers know to ready extra supply for the "TV pick-up," when households across the country simultaneously turn on the tea kettle while watching television. For the South American dam, the magic hour is connected to telenovelas. On one tour, an Itaipú engineer guide even explained, "At 7 p.m. or whenever the novela comes on, the engineers in the control room begin to generate more." Permanent power lines predetermine the users of renewable energy. Like railways, the built infrastructure of transmission lines sets the possible destinations of the electricity, but human tastes orient the ends for that electricity. Thus, the work of the energy sector takes on an anthropological quality involving guesswork about human behavior. A prediction about human predilection orchestrates the moment when water that has been stored up in the reservoir over the past day, week, or month is released.

Above-average amounts of rainfall beginning in the 1970s have contributed directly to Itaipú's ability to set generation records.[17] Climate models for the twenty-first century predict increased precipitation during the wet season and drier dry seasons than the historical mean for the twentieth century because of anthropogenic climate change.[18] According to published studies, the Paraná River basin near Itaipú is expected to avoid the devastating droughts that affected water levels in southeast Brazil in 2014. However, the modeled changes introduce two direct risks to Itaipú. Unlike Three Gorges in China, Itaipú Dam was not constructed to regulate floods; flood control competes with electricity generation and is one of the reasons the Chinese dam has less steady production. Flooding also increases siltation in the reservoir and therefore cavitation in the machinery, potentially shortening the life span of the dam even more than its predicted 150–200 years. Itaipú's reservoir lies in a fortunate location that shields it from some of the dangers of erosion, unlike the dams farther north on the Paraná. Because the binational dam was built in a rocky canyon, flooding produces minimal siltation, but upriver flooding can send sedimentation down with the current. And flooding has a direct, disastrous effect on riverine communities. In recent years, hundreds of thousands of Paraguayans (the three largest population centers of the country, Asunción, Encarnación, and Ciudad del Este, sit on rivers) have been temporarily displaced because of annual flooding with no sign of relief ahead.

Renewability implies a commitment to and dependence on place that fossil fuels do not as a basic requirement for both extraction and consumption. Interestingly, the conceptual abstractness of electricity is juxtaposed with the analogical qualities of renewable energy. To make a megawatt hour, engineers have designed a system that closely follows the contours of the earth's surface; hydroelectric infrastructure mimics geologic features, for example, lake-like reservoirs or river-like power lines, hence Itaipú is a natural machine. But even solar and wind have analogical qualities; they operate by being acted on directly by natural movements not mediated by humans. Apologists for hydroelectricity hail it as "renewable" not because it is permanent, but because extraction does not deplete the resource (all dams have operative life spans after which river siltation and mechanical fatigue render them inoperative)—the Itaipú reservoir is refilled by natural processes. In this sense, solar, wind, and hydroelectricity have more in common with canals than they do extractive mining or even commercial agriculture.

Because hydrocarbons can be stored and thus moved from away from their source, their production can be separated from consumption, a rupture that fetishizes fossil fuel energy as something materially and temporally flat. But the simultaneity of renewable energy production/consumption highlights the dyad character of hydroelectricity, impacting the specific political-economic valences of renewable energy. The temporality of renewable energy forecloses some and facilitates other political-economic options. In the case of South America, the placed-basedness of hydroelectric sources has augmented the control of local state actors, rather than that of private corporations. By the same token, it has also turned those who manage electricity into local state actors. All Itaipú infrastructure—physical, legal, and economic—follows in the precedent established by the dual culture/nature of a megawatt hour, negotiating mechanical requirements with sociopolitical ones. And those who were most aware of these qualities were the energy managers who worked in Itaipú, who daily saw how the physical infrastructure of turbines and transmission lines overlapped with human proclivities.

Making Energy National: Constructing Sovereignty

Although Itaipú promised to flood Guairá Falls, set as the border following the devastating nineteenth-century War of the Triple Alliance, memory of the war remained, and the dam merely transubstantiated the contested border into electricity. A hydroelectric dam presented new difficulties in determining national boundaries. Water in a reservoir is indivisible; electricity is less humanly tangible than wood or gasoline or charcoal; electric lines are fixed infrastructure, unmoving, and, at an instant, although they do not visibly change, they become animated as incorporeal charge "flows" "through" them. Yet the national histories behind Itaipú made boundary making ever urgent, especially within Paraguay.

Major engineering decisions made during construction (1974–91) responded as much to political crises, national and international, as they did to the limits of hydrology. The resulting technolegal formations have been set into the bedrock of the earth and of international law, poured into the concrete foundation of the hydroelectric dam and the treaty apparatus that accompany the high-tension power lines. I am not arguing that disputes over blueprints were metaphors or analogues for political struggle. They *were* political struggle. And because they were, the technical qualities of hydroelectricity at the magnitude of Itaipú form the material foundation for a new kind of political territoriality. If the geobody, the land and water, should be a notable quality (a necessary but not sufficient one) for a nation-state, then the puzzle to be solved in Itaipú was how to politicize the qualities of electricity mentioned in the previous section.[19]

Designed as a juridically distinct space of exception, the binational dam became a state within a state, simultaneously subordinate and exterior to the Brazilian and Paraguayan states. The most critical decisions were hundred-million-dollar technical ones made by individuals with electrical, civil, and mechanical engineering backgrounds headed by a twelve-member presidentially appointed Executive Directorate (six appointed by the Brazilian magnate, six by the Paraguayan), a binational board of directors with five directors from each "side" of the dam who were responsible for the areas of finance, engineering, legal, coordination, and administration who oversaw the entire construction. Two "general" or "executive" directors (the *director geral brasileiro* and the *director general paraguayo*) functioned as co-CEOs, running the Brazilian and Paraguayan operations with some autonomy but meeting bimonthly in person at the dam and speaking almost daily. In spite of the language of binationality and equality between the two countries, a hierarchy between Brazil and Paraguay was structured into the very directorate itself by the modifier "executive." For each of the other five pairs of directors, the Itaipú Treaty designated one as "executive." The executive finance director had more authority in setting financial policy and making decisions than did the finance director. The executive engineering director, likewise, had more say in engineering decisions. Both executive finance and engineering directors were contractually always Brazilian, a way for Brazil to ensure from the beginning that the dam operated at a standard on par with its already world-renowned hydroelectric energy sector. The less mission-critical roles of the executive legal and administration and coordination directors were assigned to Paraguay.

Originally, the distinction extended even to the general directors—the Brazilian general director was labeled "executive general" director while the Paraguayan was merely "general" director, but by the early 1980s, the positions were equalized, in part to recognize the administrative and engineering competence of Paraguay's first general director, Enzo Debernardi. Though the Brazilian half of the board was led by military personnel for the first two decades, the Paraguayan side was headed by a civilian engineer, even in the midst of a military regime. For Stroessner's military government in Paraguay, the point of intersection was the Colorado Party. Employment in the upper ranks of the dam and promotion in the military were both conditioned on party membership, which also controlled the presidency from the late 1940s until Lugo's election. In design and execution, Debernardi and other Itaipú technicians had to find ways to make energy generated from the dam notably Paraguayan, distinguished from the Brazilian production. They did so through the physical infrastructure.

The intellectual-legal work done by Paraguayan engineers and energy managers as they built the dam changed what it meant to be an engineer in Paraguay, even as the dam changed the physical landscape of the Paraná basin. Chandra Mukerji's labor history of the Canal du Midi (France) has uncovered an "epistemic culture of working" whereby the process of constructing infrastructure was an episteme-shifting experience for participants as collective labor drew on and fostered collective intelligence.[20] The work of building the canal involved many separate spheres of knowledge; in putting them together, a new way of thinking modern France emerged. Over time, the work of energy production has become complicated political thinking in Paraguay. The timing of the kind of labor involved in construction and operation also played a crucial role in politicizing engineering. According to Timothy Mitchell's work on the political valences of labor regimes in carbon energy industries, ongoing need for unskilled labor in coal extraction gave strength to popular sectors in the late nineteenth and early twentieth centuries, but oil, primarily needing only skilled labor for extraction, undercut organized labor and increased the power of elites over the twentieth century.[21] Because Itaipú followed the labor pattern of oil—unskilled and skilled labor required for construction, skilled labor for electricity production—hydroelectricity consolidated power for political elites.

Yet the cultural history of the dam within Paraguay inserted insistently popular concerns into hydroelectric energy extraction. Memory of the War of the Triple Alliance gave popular groups an idiom of participation, though they were not engineering experts, one that could and did affect construction decisions. Even before the 1973 treaty was signed, groups and individuals across the political spectrum in Paraguay, from the opposition to regime-supporting Colorado Party faithful, volubly demanded "defense of Paraguayan sovereignty" in Itaipú. In fact, during the brutal right-wing Stroessner government, Itaipú was one of the most popular options for public disagreement in spite of the human rights abuses that frequently followed. The Archive of Terror, for example, contains pages of confessions and surveillance reports from 1965 student protests against the Brazilian troops stationed near Guairá Falls. At first glance, it may appear that the rhetoric and target of Itaipú-oriented demonstrations were a question of defending Paraguay against Brazilian imperial incursion. But an equally important objective was to show how Paraguay's elites were beholden to elite interests in Argentina or Brazil, a way to express the illegitimacy of the Stroessner regime. And while Brazil might not have as much to fear regarding Paraguayan attacks on Brazilian sovereignty, government officials there nevertheless described decisions in terms of national priorities.

Trouble started even before the ink on the treaty dried. Treaties, though signed by executive branches, must pass legislative approval (thereby implying greater consensus). Both congressional chambers of Brazil and Paraguay fiercely debated the merits and flaws of the proposed Itaipú agreement in 1973 before passing it. Although Balmelli's Executive Directorate saw hydraulic processes outside direct human control as protecting a natural-resource-based sovereignty and though I have made a point about Itaipú as a financial-legal exception, at the time of Itaipú construction Paraguay was nevertheless a model case for the role of state violence imposed via states of exception as instrumental to state sovereignty. Agamben omitted Stroessner or Paraguay when he described the trend in modern governance to turn to the emergency to impose a state of exception that then gets normalized.

Yet General Alfredo Stroessner ruled through a permanent state of siege that suspended the constitution and criminalized protest. The state of siege was renewed every ninety days from 1954 to 1987 and lifted only once every five years for a single day to allow for elections, which Stroessner always won, before being reinstated the following day. In spite of the danger, the opposition to the government resolutely critiqued the regime's crown jewel. Indeed, there was a long-standing pattern in Paraguay for the political opposition to use technical assessment of hydroelectric projects as an antipolitical, nonpartisan way to obliquely challenge the regime.[22] The Center for Engineering Students risked the ire of the Stroessner government by their open letter. Similar acts later in the decade resulted in detentions or exile (or worse).

Among many engineering controversies, two contested design interventions in Itaipú defended sovereignty by engineering nationality into Itaipú energy, namely: (a) the locations of the turbines and of high-tension power lines, (b) and, especially, the frequency (the *ciclaje*) of generated electricity. If in the previous section, I made claims about the impact of water on the functioning of the power plant, here I emphasize the way that national sovereignty is such a central technical concern in renewable energy engineering. Argentina obstinately lurked in the shadows of the negotiations, concerned about Brazilian hegemony because cooperation between Brazil and Paraguay threatened the balance of power in South America as upriver activities affected those downriver countries. The construction of the dam overlapped with an uneasy period in Latin American international relations as continent-wide foreign policy shifted from bellicose national security doctrine scenarios to a new era of cooperation, militarily and economically. Itaipú provided the opportunities to forge a new kind of international politics.

TURBINES

Under a section labeled "Turbines," the Center for Engineering Students letter began by quoting the first Paraguayan executive director of Itaipú, Enzo Debernardi, who also helmed ANDE, the national utility. They wrote: "Engineer Debernardi has affirmed that, 'The channel of the Paraná will pass right through the middle of the powerhouse. Consequently, seven turbines will be located in Paraguayan jurisdiction and seven in Brazilian jurisdiction.'"[23] (Initial blueprints called for the dam to have a total of only fourteen turbines, a number changed several times over the next two decades.) But the students immediately expressed skepticism, despite Debernardi's assurances, by pointing to a contingency in the treaty: "The works detailed in this annex may be modified or added to, including in their amounts and means, for technical exigencies that are verified during their execution."[24] From the treaty, annexes, and public statements made by engineers from both Paraguay and Brazil, the students concluded that "Paraguay does not know how many turbines will be in its territory, such that it is possible that they will not be and that the powerhouse will be in Brazilian territory," and that even if project details established that "seven turbines would be in Paraguayan territory, this does not mean that afterward 'for technical exigencies' this will not be modified."[25] Written and spoken promises from the leadership of Itaipú did not mollify the opposition engineering students in 1973 because they understood the leeway within the phrase "technical exigency."

Years later, the matter still had traction in Paraguay. After great personal disquiet, Paraguayan lawyer Angel Rafael Céspedes finally managed to meet with Enzo Debernardi at the engineer's office. Céspedes later wrote about their 1978 exchange in *Analytic Process to Safeguard the Sovereignty of Paraguay in Itaipú* (Proceso Analítico para Salvaguardar la Soberanía del Paraguay en Itaipú), a broader legal analysis of the Itaipú Treaty.[26]

"Mr. Céspedes," said Debernardi a few minutes into their conversation, "You are upset because some of our turbines will be placed in Brazil. Why?"

"Do you wish me to be sincere?" asked Céspedes.

"Of course, say what you feel."

"Not only some of our turbines, but all of them are in Brazil," replied Céspedes.

In that dialogue, Céspedes merely echoed what others more critical of the administration had said for years: the physical location of the turbines connoted ownership. The concern in Paraguay seemed to be that if there was not a way to show clearly what Paraguay's energy was, in the future, what was to prevent Brazil from claiming all of it? But, since electricity is fungible, the question was how to locate a distinction between a quantity. Ambiguous borders, no matter the pledges of goodwill, inspired great unease. Debernardi's meeting with Céspedes five years after the student letter demonstrated the intractability of the issue. The Itaipú Directorate decided to put a definitive end to the boundary dispute that set in motion the decision to flood Guairá Falls. After excavating a temporary diversion channel to reroute the Paraná, the riverbed was exposed in 1978. As tons of poured concrete was assembled to build the retaining wall and hollow gravity dam shafts two hundred meters high, on the dry riverbed below, a line of concrete was drawn, permanently marking the division between Brazil and Paraguay. Thus, the logic went, there would be no more ambiguity about the physical border.

The directorate chose to define a physical line in the middle of the Paraná between the two countries and, as Debernardi had guaranteed, it placed all "Paraguayan" turbines west of the line and all "Brazilian" turbines east of it. The final touch had to do with where the high-tension lines leading from the turbines to the Paraguayan and Brazilian substations were placed. Even though the market was Brazilian, all lines leading from Paraguayan turbines first had to pass through Paraguayan dry land before then crossing over to Brazilian dry land and then conveying electricity throughout the country. If the water were shared, dry land was clearly marked national territory. Of course, the placement of lines even only momentarily touching Paraguayan land anticipated a future when that country would have enough demand to consume most or all of its half of the Itaipú energy. But, more than forty years into the project, Paraguay still only uses two turbines' worth of the electricity and, because Paraguay has only one 500 kilovolt and four 220 kilovolt lines leading from Itaipú into the national territory (Brazil has three 750 kilovolt lines, two 600 kilovolt lines, and one 500 kilovolt line), infrastructural and international obstacles remain over accessing more than that.

A marker in an artificially exposed riverbed might suffice to distinguish the national territory of Brazil from Paraguay, but the real challenge posed by Itaipú was how a national boundary might be drawn within a height gradient. Potential energy, after all, is based on the position of one body relative to the positions of other bodies. What kind of sovereignty could be expressed within the hydroelectric potential of the Paraná river? The solution was to conceptualize the water as shared, but the energy as equally divisible, even though the amounts of land to be flooded by the reservoir were not the same. In fact, the water and the hydroelectric potential of the Paraná were held "in condominium"—indivisibly co-owned—while the land underneath the water and the product of the hydroelectric potential could be divided. The national territory was scaled down to the subatomic level, to the electrons themselves, which became Brazilian and Paraguayan, even as energy governance jumped scale upward, from the national to the binational. It is this technopolitical arrangement that may be most suggestive in other situations of shared renewable energy resources, including solar, wind, and tidal. And so the Itaipú turbines themselves marked the beginning of a new type of national territory, and the circulating charge produced by them moving through high-tension lines across national boundaries gave rise to a different kind of national property than land or a tangible product.

HERTZ

But, if the placement of turbines elicited concerned conversations, it paled compared to the question of the frequency of those turbines, which sparked a conflict with geostrategic significance and unleashed a human rights reign of terror. Enzo Debernardi and Senator Domingo Laíno made Paraguayan history on the night of April 1, 1977. A crowd of more than a thousand pressed into an Asunción stadium to hear the Paraguayan executive director of Itaipú and his team of engineers and their opponents, a Liberal Party senator flanked by representatives from the Union of Paraguayan Industrialists, discuss the seemingly banal technical question of cycles per second. And across the country, viewers huddled around television sets, watching the first political debate ever to be televised in Paraguay with a combination of interest and dread. As leader of the country's electricity sector, Debernardi argued the merits of changing the entire electrical grid of Paraguay from fifty hertz to sixty hertz. Laíno and the Paraguayan industry leaders countered that to do so was a capitulation to Brazil that damaged Paraguayan interests. Unlikely allies, a "radical" senator and Paraguay's leading capitalists pointed out that changing the frequency would involve changing all the electric machinery of the country at great expense. The calm at the end of that April evening only masked the storm of violent repression that would follow just weeks later.[27]

In 1977, as in the United States and Canada, the Brazilian energy grid operated at a frequency of sixty hertz. The Paraguayan system, on the other hand, used fifty hertz, as did Argentina and much of Europe. (Hertz is the number of times per second alternating current [AC] polarity switches.) Recall that Brazilian Traction, Light and Power Company was started by Canadian investors; Argentina was initially electrified by European companies; and Paraguay's electrical system was set up by a company based in Argentina. This presented a problem in the technical planning for the large dam: turbines are built to produce electricity at a specific frequency. All the infrastructure—conductors, transformers—is typically built to function at one frequency. Higher-frequency equipment is usually smaller and lighter than lower-frequency equipment. Because Brazil, from the very start, was expected to use most of the energy generated from Itaipú, the earliest technical sketches for the dam had all turbines generating electricity at sixty hertz, including the Paraguayan turbines. Thus, measurements of bedrock and concrete and penstock circumference had all been made assuming the sixty-hertz size requirements, not the heavier fifty hertz.

To help tempt the Paraguayan government to accept the arrangement, the government of Brazil offered $150 million in low-interest loans to offset costs associated with the inconvenience of switching out all electricity-dependent infrastructure. Rumors, later confirmed by Debernardi, circulated that Itaipú itself would finance construction of a dam on the Monday River in Paraguay, effectively a free gift to succor the deal even more. But popular opinion within Paraguay was not so easily swayed. The Debernardi-Laíno debate culminated years of disagreement within Paraguay. The letter to the House of Deputies in 1973 also contained an exposition regarding the question of cycles. If Paraguay continued using a fifty-hertz electrical grid, warned the engineering students, it would not be able to access its Itaipú electricity. But if the country's frequency were uprooted and changed to sixty hertz, all the electricity infrastructure needed to be altered, including Acaray Dam, the sole internal-to-Paraguay fifty-hertz hydroelectric dam that at that moment supplied the entire national demand. Moreover, Paraguay also would not be able to access electricity from Yacyretá and Corpus, two massive binational Argentine-Paraguayan dams still in the planning stages at the time, but slated to produce energy at fifty hertz.

"Either case," wrote the engineering students in 1973, "leaves Paraguay damaged. We must demand that they acquire double frequency turbines."[28] Double frequency turbines, able to generate electricity at either fifty or sixty hertz, added significantly more to growing construction costs. The students' demand assumed that Brazil would be the main market but left the possibility open for future consumption by Paraguay, at least in principle. Changing Paraguay to and/or having Itaipú only produce at sixty hertz would also preclude other markets for Itaipú energy in the future—neighbors Argentina and Bolivia as well as nearby Chile and Uruguay also used fifty hertz. Only Brazil, as much of the continent used fifty hertz, could purchase Itaipú electricity without the addition of costly infrastructure. The hertz question was, in part, a question of which integration Paraguay would prioritize.

In the first half of 1977, *Criterio*, a radical Paraguayan literary magazine closely associated with the student-led opposition Independent Movement (MI, Movimiento Independiente), ran essays on Paraguayan independence in the midst of encroachment by its neighbors and detailed engineering critiques of the frequency dilemma. By this time, the decision had become urgent, as construction consortiums preparing bids needed to know the basic engineering specifications. The authoritarian regime of Stroessner watched on as progressive dissenters usurped the rhetoric of the military government, claiming to staunchly defend Paraguay against all threat. The notions that Paraguayan energy would not be produced at a frequency used within Paraguay (even if that energy were destined for the Brazilian market) or that the entire Paraguayan grid be uprooted and replaced with equipment for the Brazilian frequency were described as damaging to Paraguayan interests and ultimately endangering its sovereignty.[29] Public pressure grew until the middle of 1977 when Stroessner "patriotically" reaffirmed that changing the Paraguayan system from fifty hertz to sixty hertz was out of the question. But having just altered the constitution to allow for perpetual reelection of the president, the government brooked no dissent. Just as Stroessner effectively punted the resolution of the frequency issue to Itaipú by insisting on fifty hertz, a net of violence descended on the opposition.

Criterio was shut down by the government and its leaders apprehended starting July 19, 1977, in a sting to eliminate the Paraguayan intelligentsia and the Independent Movement to which many of them belonged. The magazine's writers and Independent Movement's leadership, as well as members of other student groups across Paraguay were detained, tortured, and imprisoned. Ministry of the Interior forces also raided and closed the offices of the Liberal Party, of which Domingo Laíno was a member. In the first few days after the *Criterio* arrests, more members of the opposition were arrested. Ricardo Canese, a young engineer and leader in the Independent Movement, was able to make his way to Holland after police burst into his home. From Europe, Canese continued his critique of Itaipú dam and the Stroessner regime. The Archive of Terror shows that the government received regular reports on his activities from afar. Others were not so fortunate and were detained for years. Those that could, fled into exile, but neighboring countries offered little shelter at the height of Operation Condor. Though it was officially denied at the time, under the secret "Condor" intelligence agreement, the military governments of Argentina, Brazil, Bolivia, Chile, Paraguay, and Uruguay pledged to surveil, detain, repatriate, and even assassinate each others' threats.[30] The Archive of Terror reveals how the dam served as a node for the transfer of documents, security personnel, and prisoners between Argentina, Brazil, and Paraguay while all three countries languished under military governments.[31] Condor activities even reached the United States when in 1976 exiled Chilean economist Orlando Letelier and his US citizen assistant Ronni Moffitt were killed by a car bomb in Washington, DC.

Outside of Paraguay, the frequency issue looked very different. Within Argentina, Stroessner's pronouncement was met with warm enthusiasm. At issue here was more than just a symbolic extension of Brazilian influence. Rather, with Paraguay staying at fifty hertz, the two Argentine-Paraguayan binational dams Yacyretá and Corpus (still only in planning) avoided the added costs and complexity of a two-frequency problem. It also served to deescalate rising nuclear tensions between Argentina and Brazil.[32] The Argentine foreign service moved forward on a tripartite agreement between Argentina, Brazil, and Paraguay (eventually signed in 1979), an important step between the two giants that led to the formation of the Common Market of the Southern Cone (Treaty of Asunción, 1989), arguably one of the most important political economic outcomes of a mechanical decision in the region.

But the technical problem remained unsolved. Stroessner's declaration notwithstanding, the $150 million loan offer still stood, and Brazilian government officials in the energy and foreign ministries believed that Debernardi would influence Paraguay to convert to sixty hertz. As the diversion channel neared completion, allowing for the most intense phase of Itaipú construction to commence, international engineering consortiums prepared bidding scenarios to compete for lucrative contracts to build the massive turbine generators. Behind the scenes, the Paraguayan government pressured the government of Brazil for an additional $250 million and the sale of military equipment. In August 1977, General José Costa Cavalcanti, the Brazilian executive director of Itaipú (Debernardi's counterpart) reiterated in public that the frequency issue was one "internal to Paraguay" and not a matter involving Brazil or requiring international diplomacy.[33] With some sympathy for Debernardi, he said, "I don't envy Engineer Debernardi, who wears different hats. One moment he wears the hat of the head of ANDE and speaks as such. The next he's the representative of the Government of Paraguay. And then he's the Director of Itaipú. I only have this as my hat and I only speak in the condition of Director of Itaipú."[34] In spite of the questionable claim made by a military general who headed a hydroelectric project, such verbal distinctions between Brazilian and Paraguayan government forms served to mark a difference between the two countries.

Seven months after the televised April debate, several different proposed solutions lay on the table. And then a surprise communiqué was issued by the Brazilian Ministry of Mines and Energy on November 10, 1977:

> The Ministry of Mines and Energy informs that, taking into account the studies carried out by the technical organs in its area, particularly Eletrobras, and the discussions that have been held over a long period at a technical and governmental level with Paraguay on the subject of the frequency of the Itaipú generating units, the Brazilian government has chosen the solution that, among the alternatives considered viable by the countries, best serves the national interests. This solution consists in the installation, at the Itaipú hydroelectric plant, of nine generating units operating on the frequency of 60 hertz and nine of 50 hertz.[35]

The decision was to have two entirely different sets of turbines rather than a handful of dual-frequency turbines: the Paraguayan half operating at fifty cycles per second, the Brazilian half at sixty cycles per second. Two different production designs for turbines meant greater expense, a more complicated construction, and twice as much turbine maintenance expertise needed. Moreover, the equipment for the fifty-hertz turbine generators was slightly heavier than the planned sixty-hertz units, requiring an entirely new set of structural calculations to determine the load-bearing capacity of concrete and of the rock below the dam. But it ended the debate and allowed construction to move forward.

Paraguayan officials were caught off guard by what looked like a unilateral decision but quickly recovered. Debernardi asserted that the decision "was not unilateral" and that it was "the best decision, technically."[36] Paraguayan foreign minister Alberto Nogués concurred that the decision established "the full and absolute equality of the two parties, that is, Brazil and Paraguay."[37] With Eletrobras's pronouncement, the $150 million low-interest loan and the Monday dam project fell through. Itaipú engineers, when drafting a technical compendium, described the situation as "not the least cost solution," but one with "no major technical difficulties" and one that was "politically acceptable to both countries" (see figure 1.4).[38]

Members of the US foreign service had taken an active interest in the frequency issue because General Electric and Westinghouse were in the running for construction contracts and because the issue offered a context to reframe international relations in the region. Embassy staff from the United States in Asunción and Brasilia regularly communicated with electricity sector officials in both countries. Their assessment of the decision was that the Brazilian government had wearied of Paraguayan stalling and balked at the ever-increasing financial demands made by the smaller country. The higher price to be paid by having two frequencies in one dam effectively neutralized the Paraguayan trump card.

In the end, neither American consortium received the contract for the turbines, possibly in part because of frustration with Paraguayan-American dealings, but also to maximize the number of locally based contracts, a move that strengthened the economic weight of Paraguay's new growing engineering middle class. To improve energy infrastructure within Paraguay, Enzo Debernardi's office applied for an Inter-American Development Bank (IDB) loan in 1977. In order to support Paraguay's application, the Carter administration made an unprecedented stipulation: that the country allow the Inter-American Human Rights Commission (IAHRC) to investigate reported abuses in Paraguay. Debernardi's surprise was understandable. After all, the IDB had disbursed a $33.6 million loan for rural electrification in Paraguay with no hesitation in 1975.[39] For years, the commission had received complaints of detentions, raids, and closures directed at religious institutions and trade organizations, but these were ignored by previous US administrations. The Argentine and Brazilian military governments, participants in similar activities, were well aware of and nonplussed by the state violence within Paraguay. They attempted to dissuade Paraguay from complying with the requests of a US government newly interested in human rights, but need prevailed.

The human rights team investigated denunciations in late 1977, including the "Juan Félix Bogado Gondra case," named after a key leader of the Independent Movement who was arrested among others in July 1977. "According to denouncers," wrote the IAHRC in their 1978 findings, "these detentions resulted from, among other reasons, the fear of the government of Paraguay of a reaction by this group to the question of changing the Paraguayan electrical system from 50 to 60 hertz."[40] Those interviewed by the commission directly connected abuses to the technomechanical question of electric cycles per second. No IDB loan followed; neither did Itaipú contracts. The project went to a consortium of South American firms instead; 85 percent of the turbine generator unit machinery was manufactured in Brazil.

The 1977 frequency decision left Paraguayan electricity in a form that the Paraguayan consumer could use, but with Brazil as the primary market, something else had to be done. Paraguayan electricity destined for consumption in Brazilian metropoles was first inverted from fifty hertz alternating current to direct current (in which form it traveled hundreds of miles from the dam to the center of the country) and then reinverted to sixty hertz

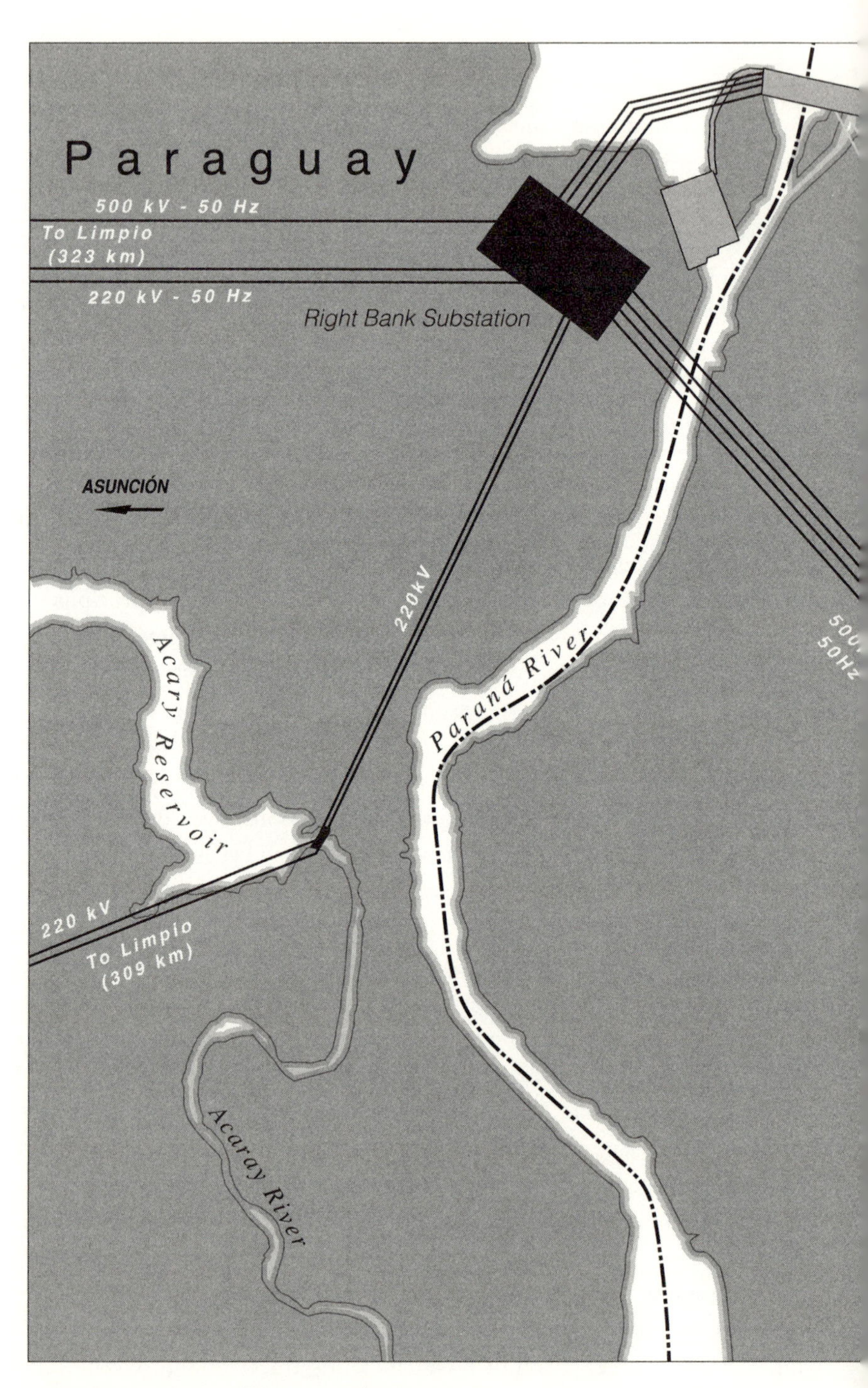

FIGURE 1.4. Itaipú transmission system diagram.
Source: Designed by the University of Wisconsin Cartography Lab.

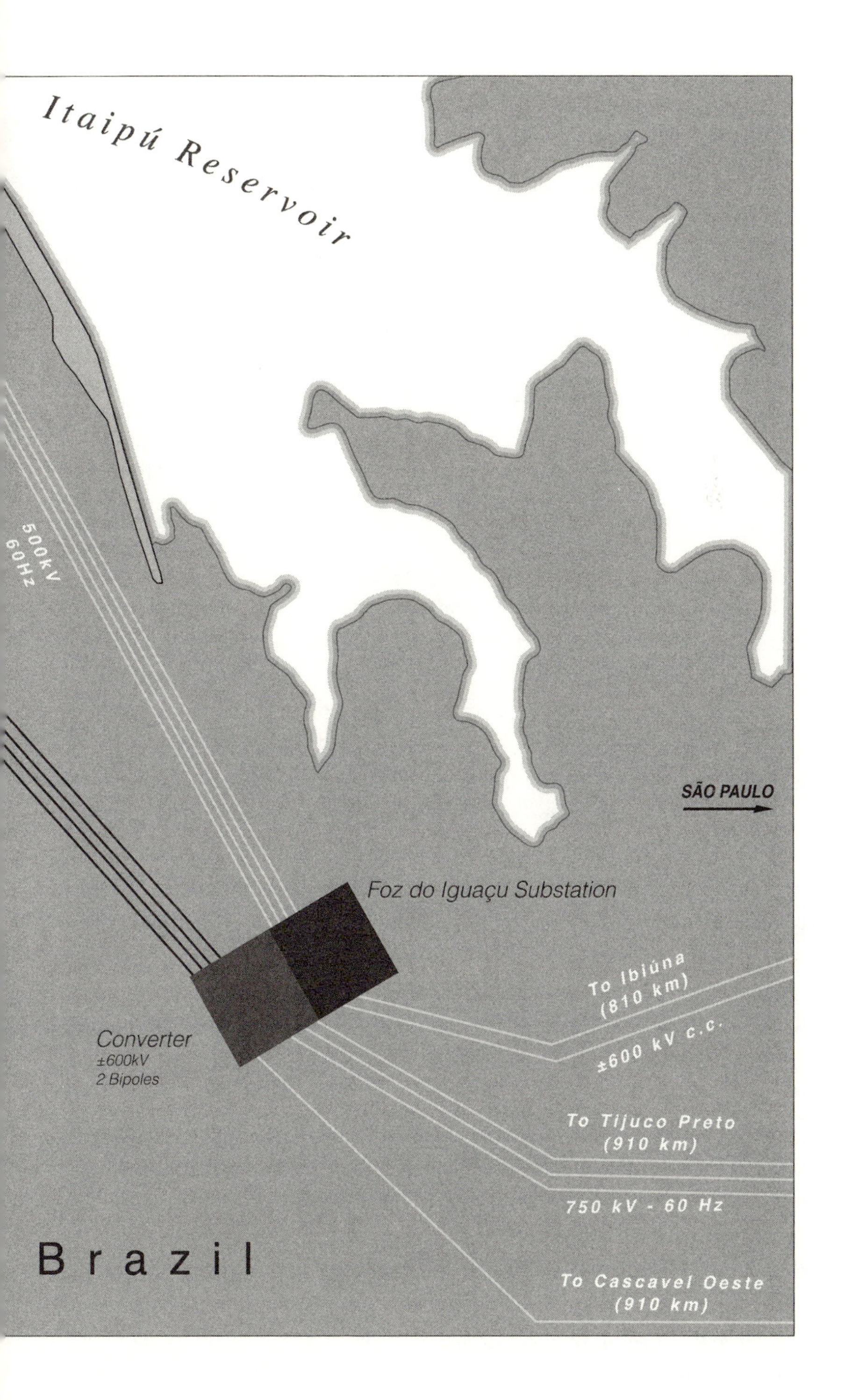
Itaipú Reservoir
500kV
60Hz
SÃO PAULO
Foz do Iguaçu Substation
Converter
±600kV
2 Bipoles
To Ibiúna
(810 km)
±600 kV c.c.
To Tijuco Preto
(910 km)
750 kV - 60 Hz
To Cascavel Oeste
(910 km)
Brazil

alternating current in São Paulo, necessitating that separate converter stations be designed and built. But one benefit was that the direct current transmission system of Itaipú served as a beta test for future long-distance hydroelectric projects in the Amazon basin.[41] All this cost more money, was more technically complicated, and lengthened the construction process; installation of turbines and energy generation were delayed by nearly two years (from 1982 to 1984). To this day, engineers in Itaipú perform maintenance on two types of turbines, which operate at slightly different rotation rates calibrated to produce about the same amount of electricity.

Just as an energy potential may become national territory, a frequency can be national and patriotic. Daniel Mains encountered something similar in his research in Ethiopia.[42] There, residents of Jimma responded positively to road construction and negatively to hydroelectric dam construction, conceiving of the latter as wasteful corruption but the former as a hopeful sign of progress. The difference lay in personal relationships to each project. Whereas the Ethiopian dam was constructed by foreigners and paid for by loans, the road was constructed by local labor and financed in part through local fund-raising drives where even the poorest individuals contributed to the national progress. Something mechanical takes on social meaning in part because of how it is treated. Because the Stroessner government responded to the frequency question as a threat to authoritarian rule, the matter became a sign of democracy within Paraguay. Because Brazilian interests desired the dam to operate at sixty hertz, a fifty-hertz frequency became an emblem of resistance to imperialism. Because the Paraguayan government insisted on its own independent frequency, Argentinian negotiators felt more comfortable with the balance of power in the region. Because the Stroessner government looked to extract more concessions from Itaipú and Brazil in exchange for changing the frequency, a more complicated, costly choice "best served the national interests [of Brazil]" by calling Paraguay's bluff.[43]

Important technological changes result in new sensibilities, aesthetic and otherwise.[44] For instance, railway travel changed not just how people read (with the concomitant invention of the mystery novel and the proliferation of newspapers), but even how people saw.[45] The natural landscape was changed not just by the dynamite that bored through mountains, but also in the vistas from train windows, blurred lines that led to impressionism. Similarly, the need to define national territory within a circulation of negatives came about when the binational dam was built. That which was invisible and/or unknown—the circulation of charge, the hydroelectric potential of the Paraná—came to be legible and meaningful and therefore the impetus for political strife. Though the turbines were tucked out of sight and no one could *see* hertz (at best, we can hear a slight difference in vibration), the political narratives kept infrastructure visible.

New political sensibilities and problems arose as the national territory was scaled down to the electron. Hydroelectricity's placedness, so necessary for energy generation, confronted political geography, the social distribution of natural resources. And so, borders of all kinds—but especially those with a bellicose past and those across wealth differentials, as are the borders between Argentina, Brazil, and Paraguay—presented a new host of problems that might be solved by engineering. Within Itaipú, the contested borders were connected to insecurities about independence and interdependence, taking on the language of sovereignty. Engineering interventions in the very physical design of Itaipú machinery defined and defended national boundaries out of concerns about how machinery itself reinforced political action. Yet since sovereignty is not a static given, but constructed, these interventions actually invented sovereignty within a binational government institution even as they defended it.

Conclusion

The technical struggles of the 1970s prefigured the conflicts in Fernando Lugo's government in several ways, including the outsized importance of engineers to national politics and engineering as a national project. Indeed, the first ever political debate to be televised in Paraguay, even under the repressive regime of Stroessner, had to do with the pros and cons of a fifty-hertz versus sixty-hertz grid. Although many of the most senior engineers have passed or retired from public life, the student activists who cut their teeth in the protests of the 1960s and 1970s formed the nucleus of Lugo's political team. Because of what Itaipú did to Paraguay in the past, engineering interventions are endowed with the hope and ability to lead to a new politics. But since electrification disproportionately benefits urban populations in Paraguay, infrastructural challenges in electricity access continue. The formal Itaipú Treaty and the informal binational decision-making customs established during the construction process set patterns and expectations for the new round of negotiations between Brazil and Paraguay in the Lugo era. In the ways that the very design of infrastructure elevates and cements priorities, energy may be thought of as both cultural production and cultural producer. And so, this chapter has showed how energy managers use engineering interventions as a way to do national and international politics.

Questions about how to make the dam most efficient were alloyed with questions of how to protect the sovereignty of Paraguay and Brazil as Itaipú engineers factored sociopolitical categories as variables into hydroelectric calculations. Even though electrons have no known components, material interventions produced nationality at the subatomic level, as if a new kind of property were tagged onto charge: Paraguayanness, Brazilianness. For all the language of cooperation and binationality, asymmetries lingered (and in some personal conversations, I heard engineers voice strong animus along national lines). The frequency debate still stands in for patriotic loyalty in Paraguay. One Paraguayan, reading and disagreeing with some statements I made in a local newspaper interview, referred to me (negatively) as "that 60Hz researcher" in the article comments, possibly assuming that I was Brazilian, but certainly implying that my loyalties were toward the larger country and against Paraguay.[46]

Mitchell has asserted that, to understand energy, we need to look beyond just the political economy of energy rent and instead consider the labor systems built for and from fossil fuel extraction. Here, I have argued further to say that, to understand energy, we need to look at the machinery used to extract and produce energy. Machinery itself bears the imprint of energy values and ethics. Reverse-engineering an Itaipú megawatt hour shows how water becomes electricity and how electrical charge acquires nationality. Like the water spun into electron flow, the engineers responsible for planning, building, and running Itaipú have themselves been transformed by the dam, converted from technicians into technocrats as the state rematerializes in energy policies and practices. Working in Itaipú and the electricity sector became an alternate source of political capital and legitimacy, both within Brazil but especially in Paraguay. Yet no former Paraguayan executive director has become president, perhaps because the technical capacities, in either engineering, law, or diplomacy, that are prerequisite for executive directorship differ from those qualities that lead to electoral victory in Paraguay.

As "power is politics," the invisible circulation of Paraguayan charge (distinct from Brazilian charge) animated conceptualizations of territory and nationality, forming the basis of a wider ethos of renewable energy. Sovereignty here comprised territory, decision-making ability, and systems requirements begun as a physical intervention, not only symbols, bodies, or law (although all these show up elsewhere). There is a curious dual reconstitution of the nation-state and the national territory in Itaipú; while the state is scaled up, transformed into binational energy governance, the national territory is scaled down. In fact, the defense of sovereignty and the invention of nationality are simultaneous assignations of territoriality. The indivisibility of moving water and the immaterial quality of electricity necessitated new methods of locating and asserting identity, territory, and ownership. In Itaipú, the infrastructure of the turbines and the very frequency of current were used to determine national identity, impacting how we may think about value when something as immaterial as charge can be made to store sociopolitical properties via interventions in the built environment. Classic political economy from Smith to Marx reminds us that value is based on social relations. That the movement of fungible charge has taken on Paraguayanness or Brazilianness without being materially altered in Itaipú underscores the social rootedness of value.

Itaipú's multiple meanings derive from the dialectic that meets in its machinery: the intention with which the infrastructure was designed (e.g., state power), the ways those interventions were read and responded to (e.g., sovereignty), requiring yet a third action. The dialectic between intention and response continues into the present and is shaped by changes in political-economic and political-ecologic context. For years, the representational relationship between Itaipú infrastructure and society resulted in state security concerns read as binational or trinational problems of sovereignty, to which the solution was to make electricity national. But with global warming and water scarcity, the meaning of Itaipú energy as threat response may be set to change once more. Uncertainty and the need to move beyond carbon make valuable water resources all the more attractive. Itaipú secures energy for future demand, revenue for governments, and a politically fraught region at a resource-rich bellicose triborder.

Hydropolitics implies a different relationship between environmental interests and economic/industrial interests than in fossil-fuel-dependent contexts. But anthropogenic climate change is increasingly a bottom-line-affecting concern for dams, and thus the antagonistic polarization between the energy industry and the climate change movement is poised to differ. Where climate change is not positioned oppositionally to the energy industry in the United States, it is often framed as a source of increased demand in the future, not as something that imperils supply. In Latin American hydroelectricity, climate change is positioned inside the industry and threatens the very logic for the entire system. That is to say, the threats faced in the twenty-first century are not just to sovereignty, but increasingly to the cyclical reliability of energy generation and of energy rent (e.g., energy security). Tellingly, both the well-established concerns around sovereignty and the growing concerns around sustainability-as-security are directly connected to the placedness of renewable energy. Its very condition of possibility is also the source of its liability.

2

Currency

Introduction

MARCH 2009

Eight months into Fernando Lugo's presidency (2008–12), dialogue had reached an impasse. The latest round of binational negotiations between Brazil and Paraguay over Paraguayan demands for more power sharing in the dam and a higher price for electricity sold in Brazil were mired amid charges that while Brazil had sought to preserve the Itaipú Treaty, Paraguay was attempting to undermine it. Possessing the moral high ground and the bargaining power gave the Brazilian negotiators the upper hand. And then a curious thing happened: math.

Often, instead of sitting at my desk on the third floor of the Asunción compound, I sat in the foyer to Carlos Mateo Balmelli's office where two, sometimes three, administrative assistants and any number of bodyguards took phone calls, arranged documents, and escorted visitors both scheduled (politicians, business leaders, foreign professionals) and unscheduled (Balmelli's advisers and other directors of the Executive Directorate Board). Guillermo Velázquez, an engineer who had worked at Yacyretá Dam and formerly as vice minister of mines and energy, was a frequent presence. In a country where party affiliation takes on factious, almost religious-ethnic connotations, Liberal Balmelli had nevertheless requested that Colorado Velázquez accompany him during his tenure as the Paraguayan executive director of Itaipú (2008–10).

One afternoon, Velázquez, with subtle but detectable self-satisfaction, handed Balmelli a green file folder containing the findings of an exercise done "por si acaso," just in case. A few days later, when I again visited the office, Balmelli happily reported that, at just the right moment, the Paraguayan directorate team had presented the news to their reticent Brazilian counterparts.

"They were in shock," he told me, replaying the silenced and confused expressions of the Brazilian leadership of Itaipú. "They said, 'No, this cannot be!' "

But Velázquez's calculations were unnerving enough that the negotiations became unstuck. Velázquez was an engineer, not an accountant. And yet, in the Paraguayan hydroelectricity sector both past and present, engineers were integral to the financial infrastructure of the dam. As part of Balmelli's close team of energy sector technocrats, he was involved in discreet negotiations with Brazilian energy sector technocrats. Though these conversations between peers who worked side by side might be less acrimonious than the more publicly visible ones led by Lugo's progressive allies in the Ministry of Foreign Relations, tensions had nevertheless arisen.

What Velázquez had done seemed innocent enough and, in fact, a little banal or even a bit "geeky," but he knew that formulas and prices could be used for social leverage.[1] Velázquez reworked all the calculations for the Itaipú energy tariff and financial obligations as stipulated according to annex C of the Itaipú Treaty from scratch. And he found that, per a little-read clause in an amended document from 1974 concerning inflation and the US dollar newly untethered from the gold standard in the midst of the global energy crisis, a case could be made that a previous calculation was in error, that a critical "inflation adjustment factor" should have been higher from the very beginning of all payments. Velázquez argued that both central governments should have been receiving more money per year according to the 1974 amendment. There was no moral high ground if neither party had strictly kept the treaty, and the international conversation lurched forward. Months later, when the Paraguayan press got wind of the discrepancy, the behind-the-scenes binational directorate negotiation was closer to the version of the agreement that would be signed by both presidents Lugo and Lula in July 2009. The lesson learned from Velázquez is that financial details wield political leverage; the lessons to be learned in the next few chapters are how the price of electricity is a moral calculation and how renewable energy money—namely energy rent and hydrodollars—works. Here I pull the curtain back, revealing the initial and annual negotiations that structured the finances of Itaipú and the ways that they reconciled competing financial goals. The price, or "tariff," of Itaipú electricity synthesizes multiple forms of energy rent, in keeping with constraints rooted in the materiality of hydroelectricity, even as it performs the political task of recognizing sovereignty and establishing ecopolitical priorities.

No subject was more controversial, no topic raised more ire, than money, the flow of that other circulation that passed through Itaipú Dam. The real question with Itaipú was not how to make energy; it was how to generate revenue. Nearly $4 billion passes through the dam every year, all of it denominated per the 1973 treaty in the neutral currency of the US dollar. Nevertheless, the finances of the dam inspired umbrage and conspiracy theories, all intensified by the bellicose national histories of Brazil and Paraguay and by the fact that the dam's ledgers were outside the legal purview of any independent accounting body or even the legislatures of either country. Justification for this came from a defense of national sovereignty, using the binationality of Itaipú as an argument that, to protect the sovereignty of both countries, neither had jurisdiction over the dam. And while no one outside the hydroelectric engineering community seemed to care much about the less-than-optimally-efficient mechanical choices made during construction, the ramifications of Itaipú finances have attracted the attention of no less prominent an economist than Jeffrey Sachs (see chapter 4). As we saw from Velázquez's pricing calculations, the complicated infrastructuring of the finances and of the energy went hand in glove and was often performed by the same individuals.

To untangle how different parts of the Brazilian and Paraguayan governments made money from Itaipú energy, I give an overview of how Itaipú finances work, highlighting decision-making processes at the executive level of the dam in order to distinguish between rent and debt as accumulation strategies. Then I sketch out scholarly approaches to energy money to argue for the interlinked histories between hydrodollars and petrodollars as well as the particularities of renewable energy financial impacts. The bulk of the chapter homes in on the technicalities of the electricity tariff, how energy rent is conceptualized within Itaipú, and its effects on the Paraguayan government. Like the "making of a megawatt hour" from the previous chapter, I describe the making of a tariff, the base cost of electricity, and the political work done by a number before turning to "compensation," the rent that Paraguay receives for energy sold in Brazil as well as the ensuing controversies that spurred the election of Fernando Lugo. I close with an ethnographic exploration of favor-petitioning practices in the Paraguayan side of the dam and how administrators in the Lugo transition attempted to curtail such expectations. These calculations cast light onto the "not-so-cool" aspects of hydrofinance that enable the entire system even as the tariff formula and the social expectations into which it is embedded reveal broader postures toward state wealth, the public, and who has a claim to the financial resources of a nation.

The Basics of Itaipú's Financial Infrastructure

In this chapter and chapter 4, I look closely at the price of electricity and the construction debt, the formulas—social and numerical—for their calculation. Both are spelled out in the Itaipú Treaty, a public document implying a financial transparency that in practice has been anything but transparent. Itaipú's legal-economic architecture was established before any engineering interventions took place. Shortly after the 1973 treaty was approved, each public utility provided $50 million as the initial capital of the dam.[2] The $100 million initial investment was nowhere near enough to pay for construction. Neither was it sufficient to secure the amount of credit necessary for the project. Large-scale hydroelectric dams are often debt financed and, because of the long delay before energy generation and thus revenue, coincide with strong state intervention in the economy. Who else could provide the collateral or support for a multibillion-dollar project with no income for more than a decade?

Formulas solve sociopolitical problems, not merely mathematical ones, by asserting equivalences and masking discrepancies; in other words, as Velázquez intuited, "formulas are important actors."[3] In addition to bringing in revenue, the equation for the electricity tariff was used to recognize and defend national sovereignty and to accomplish various political projects. How Itaipú makes money and for whom unveils moral categories at the center of renewable energy economics, that is, something else besides the market's invisible hand is involved in hydropolitics. Not only was the dam expensive, but there were tensions as Itaipú was directed by two national electricity companies. Eletrobras, the Brazilian utility and major vendor of Itaipú electricity, wielded great influence over the dam, raising questions about whether Itaipú's fiscal policies benefited the binational dam or the Brazilian electricity company. In a classic economics text, Ronald Coase characterized a "firm" (as opposed to the "market") by its suppression of the price mechanism, where the allocation of resources is determined by administrative fiat.[4] The power of administrative decision making over Itaipú finances was distributed unequally in the binational firm, with the balance of power on the Brazilian side. Thus, the energy tariff and the debt were the targets of multiple negotiation attempts: initial treaty discussions in the 1970s, new crisis-motivated renegotiations in the 1990s, and the Lugo transition negotiations of the 2000s.

At its core, the original designers of Itaipú had to reconcile two competing economic goals that appeared to be diametrically opposed: adhering to financial principles versus the profit motive. Importantly, the price of electricity was based on expenses, not the value of that electricity, or the water, or of the services provided by the dam. These are the values and priorities ("calculation grammars") that set the stage for the original design of the financing and its implementation into the present. As it was a public company, a state utility endowed with the right to steward the nation's territorial resources, the guiding principle of the dam was financial equilibrium. Since the state is not supposed to profit, Itaipú finances were to operate at cost, with no excess, no profit, no savings for the future. And yet, neither government was so altruistic.

Thus, out of the desire to build wealth and increase access to capital while breaking even came the invention of the spatial, financial, and juridical exception. The creation of Itaipú as a fully autonomous, isolated space, juridically and financially separate from the national states of Brazil and Paraguay, allowed the planners to accomplish both seemingly antagonistic goals. The binationality extended to the smallest detail, all governed by the treaty: employee salaries, the formula to calculate the tariff, debt promissory notes. From a vantage point outside the dam, the finances were anything but clear, notwithstanding the claims that everything was transparently described in the Itaipú Treaty. This led to a popular expectation within Paraguay of unbridled wealth at the disposal of fortunate and mysterious patron-client networks. Because of the national histories of Brazil and Paraguay and the territorial costs of the dam, the money in Itaipú is a moralized narrative based on sacrifice, risk, and sovereignty. Careful balancing acts in income and expenses were required to ensure sustainable financial equilibrium and the sovereignty of both countries: at times, in the accounting Itaipú was treated as a single entity; at other times, especially in rhetoric, the two halves of the dam were distinct.

While much might be said about how alleged fiscal malfeasance in Itaipú brought fabulous wealth to the so-called barons of Itaipú who constructed the dam, the legal accumulation strategies structured into the dam's finances show much more about how wealth acquisition and state power operate. Rent was the primary accumulation strategy of the Paraguayan government; compound interest, Brazil's. The very telling difference speaks to institutional arrangements and the asymmetries between Brazil and Paraguay and to the ways state power can be wielded in the modern global economy. Per the treaty, both the Paraguayan and Brazilian governments glean non-market-based rent in various forms, discreetly bundled into the price of energy. The majority of Paraguayan megawatt hours go to Brazil. But rather than receiving market-based revenue for its electricity sold in Brazil, the Paraguayan government receives a non-market based fee—another form of rent—for its megawatt hours. The compensation for energy exports has run into the hundreds of millions of dollars per year over the life span of the dam.

The Brazilian government, on the other hand, has benefitted by holding much of the dam's debt, which grew because of insufficient payments. As part of understanding how financial and political integration work, chapter 4 will explore another equation: debt plus energy equals state power. A collapsing of the creditor-debtor dyad within Itaipú was enabled by and constitutive of the state's "magical" power.[5] The amount charged for Itaipú electricity was less than the amount necessary to service the debt, an accounting irregularity initiated by one of the two co-owners of the dam, the Brazilian electricity company Eletrobras. Eletrobras officials claimed that Brazilian consumers (the main source of the dam's revenue) simply could not afford a high electricity rate. But a significant conflict of interest lay behind the assertion: the Brazilian company also happened to be the chief creditor of Itaipú. The Brazilian utility pressured the leadership of Itaipú Dam to set a price for the electricity that was below what the dam needed to charge in order to be able to pay back its own loans to the dam.

Interest compounded over the years; Itaipú's debt to Eletrobras skyrocketed by billions of dollars. In 1973, when the original plans for construction were drafted, the total debt (including interest payments over the fifty-year term of the treaty) was planned to be $2 billion. That number has surged. Today, the total debt to be paid by 2023 is $60 billion, a 3,000 percent increase. These financial practices, because they required assent from both partners, imply concessions from one country to another, raising the question of equivalences, of what is deemed a fair exchange. What is interesting and confusing is the surprising merging of identity as Eletrobras acts as owner, creditor, purchaser, and debtor. From the 1960s, Eletrobras and the broader Brazilian state have been tightly knit because under Import Substitution Industrialization the state was proactively involved in economic development and because the "executive" finance director of Itaipú, the one whose office initiated policy, was always Brazilian.

One of the striking similarities between energy and money is the importance of convertibility. If either charge or currency becomes inconvertible, if either ceases to circulate, it becomes useless. As the dam converts moving water into electricity, it also converts money into energy and vice versa; moreover, it converts both into state power. And so, both the Brazilian and Paraguayan governments found legal ways to make money off of something supposedly not revenue generating. All the transnational complexity and bureaucratic sleight of hand was condensed into a single, treaty-governed number. The price of Itaipú electricity not only covered expenses; it provided income for Brazilian and Paraguayan government institutions. Price did the political work of recognizing and protecting national sovereignty and the ecological work of properly stewarding the hydraulic natural resource of the Paraná.

Multiple institutions converge at Itaipú and participate in the circuit of payments: the dam itself, the two state utility companies, the national governments of Brazil and Paraguay, local municipalities and departments and states affected by land loss from the reservoir. Each has different constituencies and interests. And each has weathered the contraction of the state differently. Begun by the Paraguayan state, ANDE remains fully nationalized to this very day. However, Eletrobras, which was fully state owned when Itaipú started, was partially privatized in 1995 as a minority set of shares was made publicly available. That is to say, Eletrobras has split priorities—to the Brazilian nation, to its shareholders—where the profitless equilibrium principle and commitment to provide cheap energy contend with the profit motive. While it appears that binationality has shielded against privatization, Itaipú remains a labyrinth of finances and obligations, made all the more opaque because the books were kept at the dam (and not as public record or even under government control in Paraguay and Brazil). A push toward transparency in Paraguay's binational dams accompanied the Fernando Lugo government, making financial documents publicly available to an unprecedented, never repeated, degree; the personal invitation I received by the Paraguayan executive director to observe Itaipú's executive administration was also part of the new turn in Paraguayan politics.

Conceptualizing Hydrofinance versus Petrofinance

Itaipú's renewable energy financing history intertwines with the economic history of fossil fuels from the very beginning. Not only was the promised cheap hydroelectricity favorably compared to the price of petroleum, which steadily rose in the first decade of the dam's construction. But the financial fallout from hydrocarbon politics in the 1970s materially affected the finances of the dam, especially in the interest rates for loans. Yet scholarly and policy debates around the sociopolitical implications of energy finances have tended to focus on fossil fuels, even when the energy matrix is more varied (as in Latin America) or haunted by nuclear options (as in North America and Europe). The primacy of hydrocarbons in how we think energy money has led to an attention on extraction, export, and enclave making within critical scholarship in Latin America and beyond. Others set the conversation in terms of so-called resource curse arguments in Venezuela, Nigeria, and the Middle East, where a large natural-resource-based windfall disproportionately shapes politics and economics, with deleterious effects on development and democracy as other sectors of the economy lag behind and government has redistributable income sources other than a national tax base.[6] But renewable energy has different economic and political contours by nature of the materiality of the resource and as a result of the legal-historical contexts of those same resources. In the case of Itaipú in particular, the binationality and attendant integration also play crucial roles in affecting energy finances. And so, just as renewable energy engineering infrastructure takes the shape of geological features, so too does the financial architecture take the form of the state structures that construct it.

Two common framings of the impacts of energy on money are "energy rent" and "petrodollars," both taking their standard definitions implicitly or explicitly from fossil fuels. Energy rent follows in the long liberal tradition of ground rent, understood by Adam Smith as "the price paid for the use of the land" or other nonproduced inputs that are thus distinguished from riskier profits.[7] Rent seeking attempts to seize and control revenue that comes from energy extraction contribute in large part to the negative qualities often described in the petrostate literature. But in his work on Nigeria, geographer Michael Watts challenges the notion that resource curse dynamics are deformities or incomplete versions of capitalism; instead he describes those distortions as part of the complexity and contradictions within capitalism itself. Petroleum is an "ethos" in oil rentier states like Nigeria, which are guided by two "cardinal principles: how to capture oil rents and how to sow the oil revenue," and major transnational oil companies seem to adapt to these conditions quite well.[8]

The concept of petrodollars is of more recent origin, arising as analysts tried to account for the larger market dynamics associated with energy extraction, particularly in light of the energy crises of the 1970s. Though the market for oil operates in US dollars rather than national currencies (one meaning of petrodollar), the local currencies of oil-producing states are nevertheless buoyed and backed by oil prices (another common meaning of the term). This can lead to so-called Dutch Disease. Global energy prices rose over the 1960s and 1970s, coinciding with exploitation of the Netherlands' newly discovered natural gas fields, which strengthened the Dutch guilder. Dutch manufactured products became more expensive, slowing exports in favor of imports to meet national demand and resulting in an overall contraction of industry. Petrodollar is also used to describe the trade surplus created when such a valuable resource is sold by otherwise underdeveloped countries. Oil-producing economies may be unable to completely absorb the sudden influx of dollars, raising the question of where the money might be invested. Reliance on energy rent and petrodollars often results in political and economic instability. Yet scholars have argued for the need to delink energy finances from de facto capitalism or market expectations. Douglas Rogers's work on "petrobarter" in the Soviet and post-Soviet world (the exchange of oil for goods without mediation of money) shows how petrostates can have financial practices different from those described above.[9]

While sharing some of these attributes, Itaipú also displays financial dynamics that differ from those of fossil fuels, suggesting that renewable energy dependence might result in configurations that do not strictly adhere to hydrocarbon models. In many ways, Itaipú monies and energy accumulation strategies are a matter of imagination, but an imagination reined in by water. First, unlike fossil fuels, water has multiple and arguably more important uses than energy extraction. The vast majority of freshwater is used for agriculture, to a lesser degree industry, and drinking and hygiene purposes are a distant but vital third. This is why water is often considered a public good, not a commodity. In fact, economic writing on the uniqueness of water dates to antiquity; Plato wrote "for it is the rare . . . that is precious, while water is cheapest, though best."[10] Second, as discussed regarding production in chapter 1, the place-basedness of renewable energy leaves its mark: supply and demand are physically separated, and yet they are unbendingly connected via permanent infrastructure. Because resources near locations of high demand or those easier to exploit were the first to be tapped, those that remain are more distant or more difficult, perhaps even across national borders requiring physical and legal integration. Preexisting approaches to nature also play a role since, within Latin America, there is a legal expectation of state as owner of subsoil and mineral resources.

Itaipú energy rent (the focus of this chapter) shows up as both a negative and a positive political value. As we will see, it is related to patrimonialism, favor seeking, and the familiar "corruption" discourse often associated with petrostates. But energy rent is also formally structured into the electricity tariff as a beneficial social obligation between state and nation, with patrimony and social development as animating goals. The issue is not whether Itaipú energy rent is good or bad, but rather, how it is that state actors and energy managers appropriate and designate rent, to what degree this is connected to the material properties of hydroelectricity, and how energy rent constructs nature. Social science scholarship on electricity often has to do with how it is consumed and how its consumption changes communities or how electricity's "mattering" reflects human needs and desires.[11] Yet, as the single largest source of renewable energy in the hemisphere, Itaipú hydroelectricity also represents an ethos, and, in Paraguay especially, that ethos is characterized by the capture of renewable energy rent.

What I have termed Itaipú "hydrodollars"—the money made via and flowing through the dam—offer an important foil to petrodollars (see chapter 5 for more). The trade surplus inundation confronted by OPEC did not occur in the same way in Itaipú, in part because of what Andrea Ballestero calls "calculation grammars." Rather than thinking of price as the result of strict numerical formulas, grammatical thinking allows us to apprehend the social constraints, the "mathematical and moral principles" of permitted intensities among variables.[12] For example, the difference between a commodity and a public good depends on socially acceptable limits. The massive construction debt, which claims the majority of Itaipú income, and the tariff formula in general cap the possible trade surplus. Calculation grammars in hydroelectricity finances, the expectations of how money should be spent, are part of how energy is cultural production. In Paraguay, the flood of Itaipú hydrodollars transformed social groups during construction as employees of the dam received wages on par with global standards, contributing to the class power of engineers and energy managers in that country.

The two other currency dynamics expressed as petrodollars are also mirrored in Itaipú. Hydrodollars are used to trade Itaipú hydropower because the dam uses US currency. And the dam backs and even launders the national currencies of Brazil and Paraguay, stabilizing and shaping the economies through financial mechanisms built into the very price of electricity.

Making of a Tariff

The tariff, the base cost of energy charged by Itaipú, is a chimera: it has the outward appearance of being a price determined by market mechanisms, but internally it is structured by government prerogatives at the explicit behest of political projects. This, like an Itaipú megawatt hour, is where managers have been able to synthesize competing political and economic agendas—perhaps even more than in the energy units. Not only does the tariff cover the operating costs of the dam, but the governments of Brazil and Paraguay managed to include several tax-like obligations within the tariff of a supposedly tax-exempt dam. This duality of taxes-by-another-name is merely one aspect of the tariff's double duty. As in a complicated dance at a masquerade, the tariff looks like a market price because it is capped by the market price for energy in Brazil, and yet, as Itaipú is the single largest energy producer (and one that is industry-standard efficient) in that country, the Itaipú tariff actually sets the bar for the market. That is to say, the tariff allows hydroelectricity sector elites within Itaipú-Eletrobras to proactively organize the entire energy sector of Brazil (and Paraguay) and, consequently, to exert control over industry at large.

Though many things have changed, the formula for the tariff remains untouched. An equation, taken from the 1973 treaty, annex C, article III, has not been altered in the four decades that have passed:

$$\text{Cost of electricity} = \frac{[\text{Dividends} + \text{Financing of Loans} + \text{Amortization of Loans} + \text{Royalties} + \text{Administrative Fees} + \text{Costs of Exploitation (Maintenance} + \text{Wages)}]}{\div\ \text{Guaranteed Production}}$$

The Paraguayan negotiators knew that they had the greatest leverage before the final draft of the 1973 Itaipú Treaty was decided. Once formulas and procedures were signed by the foreign ministers and then passed by legislators, maneuverability would be limited to subtle interpretations of clauses. Paraguayan electricity sector experts, under the leadership of ANDE president Enzo Debernardi, met with Brazilian counterparts from the electricity sector and from the Brazilian Foreign Ministry in March 1973 to specify the financial architecture of the dam. Like Velázquez, Debernardi's expertise, too, lay in engineering.

The initial version of the financial arrangement channeled income toward the two owners of the dam, Eletrobras and ANDE. Because the dam was begun with capital from the state utility companies of Brazil and Paraguay, it owed *utilidades* to its owners, "dividends" on that initial investment. For the oversight provided by the two companies, Itaipú was also charged *resarcimiento*, "administrative fees." Thus the initial version of the tariff formula comprised wages and maintenance, debt and interest, dividends and administrative fees. Dividends and fees provided Eletrobras and ANDE with direct revenue streams from Itaipú, in addition to whatever they might derive from commercializing the energy. As the dam was to be an engine of development, with administrative fees and dividends, ANDE and Eletrobras could reinvest in the electric infrastructure of their respective countries, building capacity to consume and use the energy. (Or, they could pay lucrative wages to their leadership.)

"This first arrangement," wrote Debernardi years later, "was not sufficiently satisfying to the Paraguayan government, whose object was, as had been expressed before, not merely to not commit the country economically nor to contribute finances, but also to obtain substantial net benefits."[13] The Paraguayan government wanted more money. And so, Debernardi and his team had to look elsewhere "with great creativity" to find another source of income aside from dividends and administrative fees, one without precedent in all the other international agreements examined by the team.[14] They found it in the water. Unlike in the United States, where private property can extend below the topsoil, subsoil resources are the property of the state in Latin America. Because Itaipú used the Paraná River, argued the Paraguayan team in 1973, it therefore owed "rent" to the owners of the national territory in the form of royalties (*regalías* or *royalties*) to be paid to the two governments' treasuries. Like the legal maneuver (in chapter 1) of determining that hydroelectric potential was held in condominium, but independent national sovereignties could be demonstrated through ownership of circulating electrons, Itaipú royalties were an important technopolitical move because they created a new kind of rent.

When Debernardi presented the Brazilian delegation this newly invented category of financial obligation, the initial response was refusal and a quarrel over which country was, in fact, more underdeveloped (*subdesar-*

rollado) than the other. Only once ANDE representatives offered a number well within the bounds of the Brazilian electricity market did the Brazilian negotiators accede. Royalties to the Paraguayan and Brazil treasuries (Ministerio de Hacienda and Tesouro Nacional, respectively) also ostensibly compensated municipalities and states whose lands were flooded or affected by the creation of the reservoir. The Paraguayan and Brazilian states, at national, state, and municipal levels, were to use royalties to fund education, job creation, and new industries.

With governments forbidden from taxing Itaipú, dividends, administrative fees, and royalties were effectively levies. Though on a different scale, this is akin to Christopher Gregory's work on small microloans where many monthly fees were charged in order to keep the formal interest rate technically nonusurious.[15] Notably, all three amounts were to be evenly divided between the two sides of the dam, regardless of how the energy consumption and income generation was divided. Eletrobras and ANDE and the treasuries received equal dividends, administrative fees, and royalties. In spite of promises to redistribute Itaipú proceeds, however, much of the income lodged in the upper echelons of the two electricity sectors and among political elites. Dividends, administrative fees, and royalties brought great wealth to both governments. From 1985 to 2017, US$5,297.7 million has been transferred to the Paraguayan treasury and US$5,310.7 to the Brazilian treasury as royalties, dividends, and fees; ANDE has received US$846.7 million, and Eletrobras has received $881.6 million in total (the slight differences are apparently accounting adjustments).[16] For example, in 1991 Brazil put in place a "royalty law," dictating how money from the dam should be distributed between state, municipal, and federal governments.[17] Paraguay effectively implemented only one in 2000, raising doubts about how much was paid to dam-affected regions for the first fifteen years of income.

Operation expenses and shoehorned government obligations were the numerator in the tariff formula; the denominator was the "guaranteed energy" production of Itaipú: 75,170 gigawatt hours per year. The full costs of the dam and the full production were combined in the tariff, allowing for full financial equilibrium. In recent years, the Itaipú tariff has hovered around $42 per megawatt hour; it started as $13.70 per megawatt hour. Beginning in 1985, ANDE and Eletrobras made annual contracts to use and commercialize the energy potential of the dam (although the original treaty-specified time span for contracts had been ten years). The billing cycles, from Itaipú to the state utilities and from the state utilities to their consumers, operate on a monthly basis. Consumers in Paraguay purchase electricity from ANDE directly; Eletrobras subsidiary Furnas (which serves Brasilia, São Paulo, and Rio de Janeiro) and Copel (the Paraná state energy company) distribute the electricity in Brazil. Though Itaipú operates in US dollars, where payments are conceptually treated as a single pool of money not denominated by origin, ANDE and Eletrobras function in the respective national currencies. Paraguay's usage was expected to be less than its half and was expected to increase over time as its demand/consumption increased; from 1985 to the present, ANDE contracted to purchase 7 percent of the guaranteed production of the dam. Eletrobras agreed to purchase the remainder. This meant that ANDE paid 7 percent of the annual cost of guaranteed electricity and Eletrobras 93 percent.

But as the first turbines were tested and installed in the early 1980s, the engineers noticed something. "They are rated for seven hundred megawatts," said Henry Winters, a senior engineer with nearly four decades of experience at the dam as he gave me a two-day private lecture on the history of Itaipú. "But they can get to 740, 750 megawatts." In fact, the turbines could safely run at 760 or 770 megawatts. Former Brazilian executive director Jorge Samek reiterated as much during public presentations on Itaipú. "Our machines were made to produce 700 megawatts per unit," he said to a Montevideo gathering of Mercosur representatives in 2008. "And there are days when we come to produce 750, 760, or 770. But the average is 750 megawatts per unit."[18]

Careful planning, even more cautious oversight, and advances in technology over the 1970s and 1980s resulted in turbines that had a significantly greater capacity than the seven hundred megawatts ambitiously foreseen by the planning commission in the early 1970s. A dam operating with eighteen turbines, producing electricity at a rate of seven hundred megawatts could generate seventy-five thousand gigawatt hours per year. But with the same eighteen turbines (or eventually twenty) operating at an average of 750 megawatts, much more was possible. The year 1995 was the first in which Itaipú produced more than 75,170 gigawatt hours, and afterward generation has regularly exceeded the guaranteed amount (see back to figure 1.1 in chapter 1). Today, the expected production is in the mid-ninety-thousand gigawatt hour range, 20 percent more than just the "guaranteed energy" production. The "additional energy," because it depended on turbines running at rates higher than seven hundred megawatts and on a Paraná River flow greater than conservative calculations, could not be guaranteed. But it was planned on.

Additional energy megawatt hours have a different, a much lower, tariff than guaranteed energy. If the guaranteed tariff is about $42 per megawatt hour, the additional energy tariff sits at $5 per megawatt hour. Price does work similar to the engineering interventions we saw in chapter 1: it gives new qualities to indistinguishable electrons. Guaranteed and additional energy megawatt hours are effectively fungible. They come from the same turbines at the same time and travel down the same wires; they cannot be mechanically separated in any way. The distinctions based on engineering principles are solely visible by price. But whereas the infrastructural interventions of chapter 1 protected national sovereignty, the financial intervention of a separate tariff for "additional" energy protected the mechanical sustainability of Itaipú. Running at a higher rate adds stress to the equipment, and the idea was that the dam should be entirely solvent with just the guaranteed energy of a seven hundred megawatt supply. However, rather than penalizing the dam for riskier operations, the additional energy tariff rewarded all the financial beneficiaries of Itaipú. Only the guaranteed energy megawatts could be used to pay the costs of the dam—the debt, wages, and maintenance. Revenue from the fifteen thousand to twenty-five thousand additional gigawatt hours went exclusively to the two governments in the forms of fees and royalties (and to compensation, discussed below). In Karen Ho's *Liquidated*, Wall Street investment bankers became so accustomed to the annual bonus that they forgot that it was not part of their base salaries and could not always be relied on.[19] But, after decades of operating at an average per turbine capacity of 750 megawatts, the question in Itaipú arises as to whether this was ever a bonus or intended from the beginning.

The additional energy was initially divided under the same proportions as the guaranteed energy (7 percent and 93 percent), but in 2002 this was adjusted (to Paraguay's benefit) to 50/50, resulting in a lower average tariff for Paraguay in general. In 2017, for example, the guaranteed energy tariff was $43.80 per megawatt hour, and the additional energy was $5 per megawatt hour. The 2017 cost of an Itaipú megawatt hour to Eletrobras was $40.86 per megawatt hour; to ANDE it was $28.30 per megawatt hour.[20] Through the tariff, Eletrobras, ANDE, and the treasuries received more administrative fees and royalties, and the Brazilian directorate placated the Paraguayan directorate by granting accommodations to the Paraguayan market. But while these accommodations might have pleased the governments and energy sectors, popular dissatisfaction with the price of energy continued undeterred.

Compensation: The Price of Sovereignty

> We don't want the crumbs [migajas] that come to us from Argentina and Brazil. We want justice and what belongs to us.
>
> —BISHOP FERNANDO LUGO, SPEAKING AT THE LAUNCH OF THE "FOR OUR HYDROELECTRIC SOVEREIGNTY" CAMPAIGN IN THE MAIN CHAMBER OF PARAGUAY'S CONGRESS, DECEMBER 4, 2006.[21]

> It is an important question that many parliamentarians have asked me: if the price of the energy purchased by Brazil from Paraguay is $2.81. Oh, please. It would be an embarrassment if there were an undertaking of this kind.
>
> —BRAZILIAN EXECUTIVE DIRECTOR JORGE SAMEK TO THE PARLIAMENT OF MERCOSUR, OCTOBER 2008.[22]

MARCH 2009

"Why is the treaty 'so unfair'? Everything was for free." After months of carefully studying the history of Itaipú, the Brazilian government official was visibly exasperated with Paraguayan demands for more money. (One of the requirements for the interview was that I specifically not name him, nor name his office, though I was permitted to say that he was a government bureaucrat dedicated to studying the new negotiations with the Paraguayan government.) We sat in his office, amid maps of the Itaipú reservoir and a carefully pored-over photocopy of a book by Enzo Debernardi.

"The Paraguayan economy grew 10 percent a year while Itaipú was built," he continued in a tone of unabated outrage. "*We* are not to blame that the money that came here was not spent on Paraguay but went to the pockets of the 'barons of Itaipú' or to the regimes. It is not *our* fault that there was no industrial policy to attract industry by selling Itaipú energy for free. If they had done that, there would be industry. The final result is that Itaipú is the 'cash cow.'"

He paused the impassioned monologue just a moment to wonder, "I don't know if you can say that in English."

I looked up from my notepad and answered the implicit question, "The correct expression is 'cash cow.'"

Language lesson learned, he resumed with full vigor. "Itaipú is the cash cow, the source of income. Some people in the [original] negotiations had the idea that it would be the motor for development, but this didn't last.

From a cash cow you want to get the most cash, especially if it is from another source."

Try as they might to shift the terms of the debate on an international plane, Brazilian government officials were unable to combat the short and incisive sound bite—"Paraguay receives crumbs for its energy" (*Paraguay recibe migajas por su energía*)—I grew accustomed to hearing from friends, in print and on the radio, and from Lugo's close circle of leftist energy activists as they characterized the remuneration for Paraguayan energy sold by Eletrobras in Brazil.[23] Paraguayan displeasure with the non-market-based price ran high during the new round of Itaipú negotiations in 2008 and 2009. It was known years before the start of the hydroelectric project, even at the tense 1966 diplomatic meetings, that only Brazil had demand sufficient to match the hydroelectric potential of the Paraná. The Act of Foz do Iguaçu (1966), which terminated a year of escalating military tension, recognized Paraguayan ownership of half of the energy derived from the stretch of the Paraná between Guairá and Iguazú. And so the obvious implication was that Brazil would consume most of Paraguay's half of the energy, paying an extra amount for it. The 1966 act, while not specifying a single hydroelectric detail or other financial particularity, called for the "consuming" country to pay a "fair price" (*precio justo*), for energy "unused" by the other country.[24] But what counted as fair was left undefined.

The 1973 treaty picked up the thread by obligating an "amount necessary to compensate one Treaty Party that cedes energy to the other."[25] The language of the treaty was critical. Per the treaty and per the technical requirements of hydroelectricity, both countries agreed to purchase all the energy produced by Itaipú. This was a condition for financing—a market for all the energy had to be ready just in time with energy generation; otherwise, vendors would not extend loans to Itaipú. But, should one country not consume all its energy, it would cede (*ceder*) its energy to the other for a compensatory amount (*compensación*) that would be paid to the national treasury. Compensation, like royalties, was paid monthly by the dam. Notice that the terms here were of "cession" and "compensation," not "selling" and "payment," and the phrase "fair price" was nowhere to be found.

Thus, yet another complicated formula was structured into the Itaipú Treaty: the compensatory amount was calculated by multiplying the amount of energy ceded by $1,200 per gigawatt hour and by a nonmarket "adjustment factor" that attempted to take into account inflation and changes in the value of the US dollar.[26] This is the same adjustment factor that

Velázquez claimed should be higher; it was also part of the royalties formula. "Compensation" immediately sparked disagreement in Paraguay within the establishment and even among opposition groups as the treaty was initially debated in congress in 1973. But rather than complaining about a formula based on an "adjustment factor" pegged to the inflation of the US dollar or the language of "cede" rather than "sell"—both of which framed the compensation as a capped, non-market-based form of rent—the contentions in 1973 were about currency denomination and the price of energy in Paraguay.[27]

Itaipú charges, in effect, four different prices for electricity generated simultaneously by the same water at the same dam: the price of guaranteed energy, the price of additional energy, the price of ceded guaranteed energy, and the price of ceded additional energy. The tariff secures sovereignty and sustainability, drawing national borders, delineating jurisdiction, and distinguishing engineering qualities in order to guarantee the renewability of the energy. The differences in these prices were set as various combinations of fees in addition to the financial obligations of wages, operating costs, and debt. But at the core, compensation was a form of rent by fiat, not a market-sensitive price. For decades, the official Paraguayan government position was contentment with the arrangement; the total received in compensation from 1985 to 2017 was US$3,722.7 million.[28] In spite of liberal economic language in the fiercely anticommunist Paraguayan regime, Paraguayan negotiators averted risk by insisting on the path of least resistance, rent instead of riskier market-seeking practices, given that the Brazilian market would be the major source of Itaipú income.

By 2008, compensation from Eletrobras to Itaipú for ceded energy was around $2.80 per megawatt hour, in addition to the base tariff. Nevertheless I was repeatedly told (accurately) that the market price for electricity in Brazil ranged from $40 to $120 per megawatt hour, after paying the tariff plus compensation (about $45 per megawatt hour). Eletrobras, its subsidiaries, and its shareholders pocketed the difference after transmission costs. While Colorado Party elite might be content with the compensation formula, ordinary citizens were not. Janet Roitman has found that struggles over a "just price" for taxes and profit in the Chad basin involved contestations over what counted as sanctioned wealth and the limits of state regulatory power over economies.[29] That is, citizens on the territorial or legal margins of Cameroon questioned the right of state actors (national and colonial) to extract taxes and other forms of rent. But in Itaipú, the problems over just price were not merely what was appropriately paid to a state entity.

Rather, it was a question of what was rightly charged by one national state entity to another (i.e., by the Paraguayan government to Eletrobras).

Standing before an expectant crowd gathered at the auditorium of Paraguay's national congress, Bishop Fernando Lugo launched an apolitical, antipolitical campaign, "For Our Hydroelectric Sovereignty," in December 2006 (two years before winning the presidency), as a way to demand change within Paraguay by demanding change in its relationships with Argentina and Brazil. With leftist energy expert Ricardo Canese at his side, Lugo declared, "We don't want the crumbs that come to us from Argentina and Brazil. We want justice and what belongs to us. . . . No sovereign country would accept that its principal natural wealth be carried away at cost," which was received with warm applause.[30]

"To what crumbs does Bishop Lugo refer?" asked the writer of the *ABC Color* article describing the event: "[The] compensation, that is the $2.33 per megawatt hour, is the price that Brazil pays our country for the approximately 36,500,000 megawatt hours that we are obligated to yield every year. The average price of the Brazilian market is about $40 per megawatt hour today [i.e., $40 in addition to the tariff plus compensation]." While Colorado party elite had enjoyed the decades-long arrangement, the mass of the Paraguayan population had seen few of the benefits. *ABC Color*, the Paraguayan paper of record, positioned itself against the Colorado Party management of the dam, regularly giving favorable coverage to the liberation theology priest and his movement for hydroelectric sovereignty. Discontent over the transpartisan critique snowballed into one of the main initiatives of the Lugo regime when he came to power (see Folch 2015, and chapters 3 and 5). Instead of compensation, Lugo and Canese asked for a market-set price for Itaipú energy. "The Itaipú and Yacyretá treaties divest Paraguay of its hydroelectric sovereignty. And, because of that, it cannot freely export its electricity at the market price," wrote Canese.[31]

After Lugo's 2008 inauguration, the parliament of Mercosur convened a special session with the expressed purpose of better understanding the growing disagreement between two of their founding members. To expanding complaints that Paraguay received only crumbs for its energy, Brazilian Itaipú executive director Jorge Samek carefully explained the amount of other income obtained by the Paraguayan government. With PowerPoint in hand, Samek showed how billions of dollars in Itaipú royalties, dividends, administrative fees, and social fund projects had been distributed to both central governments from 1985 to the 2008 date of the presentation as well as the advantageous industry-stimulating effects of inexpensive electricity.

In other words, like Velázquez, Samek went into financial details in order to connect monetary values to social values, assuming that the former were legally persuasive data for the Parlasur audience. His defense of the Brazilian government position and the status quo drew attention from parliamentarians not from Brazil. One representative from Uruguay took the floor to say, "We are bold to say that the request from Paraguay is correct and legitimate. We have to say what our opinion is, as a country, and we ought to do so because I believe it is just."[32] Without punitive power, Parlasur could do little more than serve as an international public platform to amplify Paraguayan complaints.

Because of exposés regularly published by the major Paraguayan newspapers, ordinary Paraguayans easily commanded basic facts about Itaipú electricity. It was public knowledge, for example, that Brazil exported electricity to Uruguay at higher rates than Paraguay exported to Brazil. And when the Uruguayan paper *Búsqueda* published a 2009 article about the "catastrophic" energy situation in the country, Paraguayan negotiators in the Itaipú Directorate leapt to take advantage of it. In anticipation of one of the regular binational directorate negotiation meetings, Balmelli's administrative staff assembled folders for every director. Diego Ferrer, the tech whiz who had risen to his administrative post from his start as the shoeshine boy in the lobby, made sure to print enough *Búsqueda* articles for each folder. With a fluorescent yellow marker, he highlighted a key section: "the cost per megawatt hour we imported from Brazil was $188."[33] Hence arose the Paraguayan complaint that the country received mere crumbs in spite of Brazilian executive director Samek's assertion that Paraguay also received royalties, dividends, administrative fees, and wages.

Consumers in Brazil and Paraguay saw only averages, of course, averages that hid political-moral calculations. The "guaranteed energy" tariff formula summed government fees, the construction debt, and the cost of operation; the "additional energy" formula sanctioned the environmental and engineering risk of operating at higher than originally stipulated rates; the compensation formula recognized the nationality of energy. All these were determined in annual board meetings between the binational Itaipú Executive Directorate and the twelve-person Itaipú Governing Council (which included ANDE and Eletrobras presidents as well as representatives from both foreign ministries), and, because Itaipú was such an important energy provider, these calculations set the standard of both the Paraguayan and Brazilian electricity markets.

Rent as Accumulation Strategy

The dam finally began paying obligations other than employee salaries and material costs in 1985. Although the first turbines generated electricity in 1984, energy sales commenced the following year, providing revenue at last, allowing planners to implement the financial components of the tariff formula. Between 1985 and 2007, the total paid to both governments in the form of royalties was US$6 billion.[34] In 2008—the first year of Fernando Lugo's presidency and the regime's pledge to change hydroelectric politics—each executive director had as his responsibility an annual budget of $307 million out of which to pay salaries, operations, and maintenance, and to fund social programs. Paraguay's treasury also received $117 million in 2008 and $107 million in 2009 in compensation for energy exported to Brazil. The difference in these figures depended on differences in energy sales from those years. This was added to the royalties received by both government treasuries ($240 million in 2008; $219 million in 2009), and so the total paid to the Paraguayan treasury by Itaipú was $357 million in 2008 and $326 million in 2009. For the same years, Itaipú paid ANDE and Eletrobras each $46 million (2008) and $39 million (2009) in administrative fees and dividends. (In 2008, Itaipú also paid $2.064 billion toward its construction debt.) The impact of Itaipú on the Paraguayan economy, through money to the treasury, through fees and dividends and energy sales on the Paraguayan market to ANDE, and the Itaipú Paraguay operating budget was nearly $1 billion annually in a country with a 2008 GDP of $16 billion. The income from royalties and cession of energy flooded the Paraguayan government with liquidity (see table 2.1).

In Paraguay, the increased government revenue from Itaipú served as an alternative to taxes, underwriting government projects and daily expenses as the state apparatus grew in size and complexity, with little strain (or accountability) to the citizenry. Three different public entities in Paraguay handled income from Itaipú: the Paraguayan treasury under the minister of the treasury, ANDE under the president of ANDE, and Itaipú under the Paraguayan executive director. With no congressional oversight, Itaipú-generated money could be and was spent at the personal discretion of ANDE's and Eletrobras's presidents, the finance ministers, and executive directors. At the time of Itaipú's construction, the presidency of ANDE, the directorships of Itaipú, and the finance minister on the Paraguayan side were presidential appointments, answerable to Alfredo Stroessner personally.

TABLE 2.1. Itaipú Payments in US$ Millions, 1985–2017

	1985–2016	2017	Total
PARAGUAYAN GOVERNMENT	8,403.5	616.9	9,020.4
Royalties	5,039.8	257.9	5,297.7
Compensation	3,363.7	359.0	3,722.7
ANDE	803.5	43.2	846.7
Dividends	411.7	23.5	435.2
Administrative fees	391.8	19.7	411.5
PARAGUAY SUBTOTAL	9,207.0	660.1	9,867.1
BRAZILIAN GOVERNMENT	5,310.7	255.8	5,566.5
Royalties	5,310.7	255.8	5,566.5
ELETROBRAS	838.4	43.2	881.6
Dividends	433.7	23.5	457.2
Administrative fees	404.7	19.7	424.4
BRAZIL SUBTOTAL	6,149.1	299.0	6,448.1
TOTAL	15,356.1	959.1	16,315.2

Source: Itaipú 2018:45.

And until Fernando Lugo's election in 2008, these positions were in the hands of the Colorado Party.

"Itaipú pays Paraguay royalties and compensation," said Hugo Mesa as we met in his office at the Palacio de los López (the Paraguayan equivalent of the White House) in November 2009. Mesa, with master's-level training in economics and international relations, served as an adviser to the Ministry of Foreign Relations Negotiating Team and to President Lugo himself. "I believe this has generated a deformity," he said. "The treasury has the [full] information on this, but I believe there are municipalities where royalties provide half of the budget for the municipality. The municipality isn't self-sustaining. This is fiscal negligence [*pereza fiscal*]."

Paraguay's status as a rentier state was merely bolstered by Itaipú monies. As early as 1975, Paraguayan economist Agustin Oscar Flecha had cautioned against dependency and underdevelopment.[35] Because Paraguay had to recur to external financing for unusual expenditures, Flecha urged that the short-term benefits of the economic boost from Itaipú construction be distributed equitably and intentionally within Paraguay in order to accelerate development. As the Brazilian government official alluded, Flecha thought Itaipú had the potential to develop Paraguay if backward linkages associated with construction and energy sales were deliberately reinvested in economic development. But four decades later, promises remained undelivered. The Center for Analysis and Diffusion of the Paraguayan Economy (CADEP), a nonpartisan economic think tank in Asunción, concurred with Mesa's back-of-the-envelope calculations.

Working with numbers published by the treasury and the National Bank, CADEP found that over the first decade of the royalties distribution law (2000–2010), dependence on "energy-derived financial transfers" from the central government to departments and municipalities grew an average of 21.6 percent a year.[36] Departments received $20.5 million in 2010; municipalities got $85.3 million that same year, most of which was spent on capital projects like road improvement, and very little to none of which was spent on education or health.[37] These numbers, nearly 40 percent of municipal income, represent Itaipú royalties and compensation for flooded land paid by Yacyretá Dam (the Argentine-Paraguayan hydroelectric dam). The vast majority of energy-derived financing—92 percent—came from Itaipú royalties alone; the Argentine-Paraguayan dam accounted for 8 percent.[38]

As Itaipú royalty transfers to municipal and department governments increased in the twenty-first century, Itaipú compensation payments to the Paraguayan central government increased. To many in the unwieldy Lugo coalition administration, the degree of government dependence on Itaipú was perceived as a problem. But for all the concern, rent is a hard habit to break. Over the first two years of Lugo's presidency (2008–10), hydroelectric energy financial transfers to municipalities and departments increased by 40 percent as the royalty law went into greater effect.[39] And when Balmelli was replaced by Lugo's close ally Gustavo Codas as interim executive director, they grew all the more, doubling from $105.8 million in 2010 to $237.2 million in 2012.[40]

The impression of Itaipú as a source of riches accessible to those fortunate enough to be situated within the right network of relationships extended beyond the Paraguayan government. It characterized popular perceptions. In fact, as soon as I began research within the dam, to my consternation people started asking *me* for help. One woman asked me to see about an internship for her son: "Tell them that I am a single mother, taking care of my one child all by myself," she said, before adding, "Cris, why don't you try to get a job there? They have so much money." Some requested help finding employment after a series of targeted questions ("And to whom do you talk to when you go to Itaipú?"), and another woman saying, "It has always been my dream to get into one of the binational dams." And when I demurred and said, "I don't think that's possible, but I will see if an opportunity comes up" (my least-awkward attempt to say "no" politely), no one seemed offended.

Fielding favor requests was a regular part of working in the Executive Directorate of Itaipú. Employees tried to avoid phone calls that they knew would be requests to put in a good word. If ordinary employees (and anthropology PhD candidates) received phone calls from acquaintances, it was all the more exacerbated for the new Liberal party executive director Carlos Mateo Balmelli—a public figure and not just a functionary. What differentiated these from business proposals and made them favors was the personalistic logic that explicitly sidestepped expected bureaucratic mechanisms (such as contests) and/or that the substance of these petitions departed considerably from Itaipú's mission. At no time did I hear of anyone asking for construction of power lines to improve electricity access in a community.

After nearly six months in office, executive director Balmelli found a solution. He had two versions of a stamp designed, saying "Pay for it out of your own pocket" ("Pague de su bolsillo" and "Paguen de sus bolsillos")—singular and plural; notably using the formal, more distant second-person singular *usted* form, which is unusual in Paraguay; *vos* is more common. When a request for funding made its way into his hands, he gleefully signed and stamped it on the bottom with his official seal (his full name and title) and then, at the very top of the document, stamped it with "Pague de su bolsillo" and initialized it, to complete the formality. News of this was leaked (certainly by Balmelli's office) to the media immediately. The backlash was instant: some praised Balmelli's willingness to stand up against the requests; many others excoriated him for being insolent. One exasperated Liberal Party activist even said to me, "When the Colorados were in power, I could get things, and now, when we're in power, I get nothing?"

But in the end, neither Balmelli nor Codas was able to deliver Itaipú-Paraguay from its past. News of clientelistic favors within the Paraguayan side of the dam appear regularly in the national press. Compensation and royalties are still, as best as can be determined, used to cover recurring costs or inadequately accountable education and public works projects, rather than targeted on growth-oriented investment in job creation. And if Mesa was concerned about fiscal negligence, it has continued unabated after the Lugo presidency. Today, the numbers are even more stark: in 2016, of the US$250 million transferred by the central government to municipalities in Paraguay, a full US$213 million (92.7 percent) came from Paraguay's binationals.[41]

Conclusion

In chapter 1 I traced the "making of a megawatt hour" to show how the material properties of hydroelectricity were knit with socioeconomic decisions in ways that gave a "national" quality to the Itaipú energy. Here I traced the "making of a tariff" to explore how money flowed between the dam and various institutions of the Brazilian and Paraguayan states. Although most Brazilians and Paraguayans received the tariff as a single number onto which transmission costs and processing fees were added in their monthly billing statements, the tariff itself summed various addends that were the end results of planning by state actors, synthesizing priorities and problems of sovereignty, time, nature, and citizenship. But, again, the binationality of the dam—which legitimated the financial exception—added layers of complexity as price was used to recognize and defend national sovereignty, resulting in multiple prices for indistinguishable units of Itaipú electricity. Price, as a moral-technoeconomic assemblage, ensured responsible engineering and patriotism. But because the dam's expenditures are measures that shroud politically determined calculations as market-set prices, financial solvency has metamorphosed into an issue of security.

National identities and national futures get vocalized through energy of all forms, but whereas the extraction of fossil fuel resources is often configured as ideally "unentangled" from the local, the price of the electricity and the form of the debt in Itaipú are intentionally attached to the local.[42] Money within a binational context takes on more than just convertibility; it becomes a sign of allegiance. The Paraguayanness and Brazilianness of Itaipú financial entanglements were celebrated from the commencement of construction to the very present because the dam was to be a "motor" for development for each country. Indeed, the promise of progress justified the sacrifice of the Guairá Falls. As the largest power plant in the world, Itaipú Dam provides an important counterexample to straightforward chronicles of unchecked neoliberal ascendancy in late twentieth-century Latin America. Neoliberal orthodoxy lauds the neutrality of finance as more fair and more efficient at equitably distributing resources than governments embroiled in archaic if not corrupt patron-client relationships. Yet Itaipú remained government owned even though the neoliberal privatization of public goods, including energy concerns and water distribution, spread across the continent.

Hugo Mesa, from his seat as an adviser to Lugo, saw a government suffering ills recognizable from the resource curse literature. His concern had much in common with the critique from the Brazilian government official: Itaipú had become a cash cow for many within the Paraguayan government. Many tariff components, not merely the royalties distributed to municipalities, functioned as energy rent. To be clear, though Itaipú became a source of rent and client-patron-based favors in Paraguay, it was not the initial cause of such practices. Debernardi's invention of hydroelectric royalties resulted from such expectations. Government appointment to positions led to the popular notion in Paraguay that positions were unmerited favors and not gained by professional expertise and thus that anyone might also try their luck at requesting them. And, in these ways, the hydrostate in Paraguay is reminiscent of petrostate weaknesses.

But to see only fiscal negligence ignores all the other political implications of the Itaipú financing we have seen: price performing a technopolitical function of distinguishing between guaranteed energy production and riskier additional energy; royalties deserved as a result of joint ownership of a hydraulic potential; and rent charged for compensation. Through the tariff obligations, Itaipú was to be a "motor" for the economic development of Brazil and Paraguay. By having a politically motivated price for energy, local industry also benefitted from state involvement in electricity financial planning. But the illiberal pricing practices of the tariff came to only a fraction of the very dubious accounting practices that led to the Itaipú debt. What we have seen in the tariff is that rent reduces risk and that the geological distribution of water entails sovereignty. Moreover, because payments from the dam are in US dollars whereas payments to the dam are in guaranies and reales, Itaipú launders and bolsters the Paraguayan and Brazilian currencies—a hydrodollar function. The required arbitrage of the US dollar in Itaipú then stabilizes government incomes, both individual and corporate, via salaries, royalties, administrative fees, dividends, and compensation (as well as via dollar-denominated debt repayment to Eletrobras until the mid-1990s).

Itaipú, where administrative fiat governs the distribution of resources in and through the tariff, is clearly a Coasian firm. Yet this nonmarket organization produces nearly 20 percent of all the electricity consumed in Brazil, suggesting that the entire Brazilian market rests on a bedrock of Coasian organizations and that, without them, the market could not exist. This, of course, is the case with energy resources in general; petroleum in the United States, for example, is heavily subsidized. The persistence of rent and sociopolitical values (calculation grammars) as the price structure mechanism for the largest energy source in the region complicates naive liberal ideals of unfettered markets. The situation presents a bind. The dam-as-firm is fundamentally necessary to the functioning of Brazilian industry and is a major financial input to the Paraguayan GDP through direct wages and transfers to the government. And, yet, the personal aspects of administrative fiat—where choices depend on the character and qualifications of individuals—necessarily impact the outcomes of decision-making processes. Complex economic systems are, by definition, complex and contain both market-driven and fiat-based accumulation strategies.

PART II
Integrations

3

Renegotiating Integration

Introduction

NOVEMBER 2008

"A bit of history: during the opposition [to Stroessner] we rallied against Itaipú," Ricardo Canese said to me before I could even ask a question.

We sat at the Asunción headquarters of Tekojoja, the popular social movement that launched former bishop Fernando Lugo's successful presidential campaign. On election day in April, the building had brimmed with community organizers and foreign press, but on the afternoon of our interview, it was almost unrecognizably quiet, just an ordinary house shrouded by trees in a residential neighborhood. Only a handful of visitors were in the lobby where I, too, had waited until Canese arrived at 3 p.m. after a day of meetings elsewhere. Canese, a lifelong progressive activist exiled in his youth as an engineering student after the hertz controversy of 1977, spearheaded the Ministry of Foreign Relations Itaipú negotiating team in Lugo's government.

"Now," he continued, "with the election of Lugo, it is a central axis. There's been a transcendental change here, and Brazil has noted this. And, for this reason, agreed to sit at the table." Lugo's close collaborators knew the prospect of renegotiating the binational integration opened the possibility for a new national politics and vice versa.

Canese spoke with a charismatic ease and no notes, quickly running through a list of demands that he and a working group of progressive lawyers, engineers, and economists had drafted over months. “The Six Points for discussion,” he said, “that we have proposed and that Brazil has agreed to discuss without preconditions are: (1) electric sovereignty or ‘unrestricted availability’ [*libre disponibilidad*], (2) fair price, (3) to pay the legitimate debt . . . , (4) equal administration between Paraguay and Brazil . . . , (5) control over loans . . . , (6) pending projects . . . that are parts of the treaty that still haven’t been executed.”

After laying out the Paraguayan goals, Canese described the state of negotiations: “Brazil is tough in negotiating, but we are also firm. But Brazil has at least accepted the agenda. [Brazilian president Lula da Silva] is favorable; he at least is open to discussion.” He likened Brazil to the United States, which took years before sitting at the table with Vietnam.

“What are arguments that you’re using?” I asked. “What has worked?”

“Electric sovereignty,” he replied. “Brazil also defends its sovereignty, regarding the Amazon and its petroleum. But they’re unwilling to recognize Paraguayan sovereignty in this matter, and so there’s a contradiction in its attitude. They have physical might, but they’re not in a juridically strong position here.” Canese returned to the subject of the Brazilian presidency to make an argument about a new continent-wide political model that counterbalanced increasing regional integration, including energy integration, with the recognition of multiple kinds of sovereignty. “Lula is sympathetic because he understands that in Latin America there’s a new model—one of sovereignty and integration,” he said. In fact, Brazilian president Lula da Silva was more than just sympathetic; he took a proactive role in the region to support progressive policies and equitable development projects proposed by other left-of-center governments. But not all agreed with Lula; resistance to left-turn and post-neoliberal economic policies at home and abroad grew simultaneously. “Obviously within [Brazil’s Foreign Ministry] and business sectors, there’s opposition to Paraguayan sovereignty. . . . We haven’t gotten to the point of taking the issue to an international court or arbitration. And we have the deadline of August 15, 2009, for this process. At that point, we’ll evaluate the strategy, where we are, and rethink this all,” said Canese.

The strategy proposed was a Foreign-Ministry-to-Foreign-Ministry renegotiation of the treaty before its 2023 expiration, relying on the leftist affinities between presidents Lula and Lugo to ensure forward movement within the chancelleries because there would be significant resistance from other quarters (as there was). And the chief goal was hydroelectric sovereignty, sometimes referred to as electric or energy sovereignty, which would mean the ability to access all of Paraguay's electricity production without any preexisting legal constraints or stipulations. But Canese's analysis of the Itaipú situation, the hemispheric political-economic context, and the negotiating team he led were not the only Paraguayan players or perspectives on the field. Indeed, another energy sector group advanced another model of integration, different stakes of sovereignty, and an alternative imagination of how state power worked and of how the mechanics of hydroelectricity fit into all this.

NOVEMBER 2008

"Something is not working. I have to find a new strategy to deal with Brazil," Itaipú executive director Carlos Mateo Balmelli said in our very first interview. After I had accompanied two hundred Liberal Party protestors into his office the day before, he had invited me back for a more extended meeting. Though I had been told that we would start at 9 a.m. and I had arrived more than thirty minutes early, we finally began two hours later. Not yet familiar with a normal day in the top floor Executive Directorate (i.e., that it involved unending crisis management), I was concerned that the delay signaled a change of heart, but my fears were allayed by an hour-long conversation followed by an invitation to visit the dam the very next day. We sat in Balmelli's corner office in the Itaipú headquarters in Asunción, an imposing room more than five hundred square feet large. He spoke to me in English with periodic switches into Spanish.

"I don't think we will renegotiate the treaty because it's sacred for Brazil. . . . Brazil is stubborn on *libre disponibilidad* [unrestricted availability], and we cannot sell to Argentina because the network has to be built," he continued. "But, we could sell it to Brazil's market directly through ANDE." Balmelli then got up from the large mahogany desk and walked to the bookshelves built into the east-facing wall of the room. He quickly chose one and returned back to his seat to continue the conversation. Opening the book, he pressed the pages flat so that I could read them and with a pen drew brackets around two paragraphs: articles XIII and XIV of the Itaipú Treaty.

> ARTICLE XIII
>
> The energy produced by the hydroelectric use to which Article I refers shall be divided in equal parts between the two countries, being recognized in each the right of acquisition, in the form established in Article

XIV, of the energy that is not used by the other country for its own consumption.

Unique paragraph: The Signatory Parties commit to acquire, together or separately in which form they agree, the total installed capacity.

ARTICLE XIV

The acquisition of the services of electricity of Itaipú shall be realized by Eletrobras and by ANDE, which also may do so by means of intermediaries of the companies or Brazilian or Paraguayan firms that they indicate.

Echoing what the Brazilian foreign minister had said in 2006 when Lugo first called for a renegotiation of Itaipú, Balmelli saw both political and legal obstacles to treaty renegotiation. Not only was it procedurally difficult for a weaker party to force the renegotiation of an onerous treaty, even were the other party amenable, but in Brazil there was little political will to do so. Treaties had to be negotiated by foreign ministries and then ratified by congress. Practically speaking, the call to renegotiate a treaty was a political no-go to Balmelli, though it did produce a patriotic sound bite. And so, depending on the good graces of a sympathetic president in Lula was not enough, particularly as Lula was concerned for his own legacy.

But as a trained lawyer, Balmelli had found a way from within the confines of the treaty to press for Paraguay's right to sell its excess energy in Brazil at the price of that market. According to article XIII, the energy from the dam was to be divided in equal parts between the two countries, which agreed to "acquire the energy" either "separately" (i.e., operating as two distinct energy companies) or "together" (i.e., operating jointly), all of which they would then sell via the state electricity companies or their selected intermediaries. At present, Itaipú's electricity was acquired by ANDE and Eletrobras separately and sold separately in each country. Yet, since this energy could also be acquired together and then sold together, it logically followed that, were they to acquire it together, they would be able to sell it together in both countries. That is, Eletrobras (paired with ANDE) would be able to sell to the Paraguayan market and ANDE (paired with Eletrobras) to the Brazilian market. From this, Balmelli argued that the treaty itself foresaw that each country's electricity company might sell electricity from Itaipú to the other. He thus intended to shift the debate from a question of renegotiating the treaty to enforcing the treaty with the goal that Paraguay's electric utility sell electricity directly on the Brazilian market for a market-based price and in this way finance the industrialization of Paraguay while building institutional capacity.

During the Fernando Lugo administration (2008–12), energy negotiations facilitated new models of integration and necessitated creative ways of addressing sovereignty. Lugo's presidency was an inflection point in Paraguay's nation-state formation because new individuals, new parties (accompanied by their own patron-client networks and epistemological assumptions about state power) were now in positions of leadership at the central government level. And through the economic and symbolic capabilities of Itaipú Dam, that is, through hydropolitics, these newer groups could actually implement their political visions. One key wrinkle in the government's ambition to redirect the nation's hydroelectric resources was the fact that the president assigned the power and responsibility to treat with Brazil to two different sets of actors—one within the Ministry of Foreign Relations, the other within Itaipú itself. The Ministry of Foreign Relations was in the hands of Tekojoja and Lugo's progressive allies, activists with years of community organizing experience and decades of work on the issue of Itaipú but less experience as the government.[1] Itaipú, on the other hand, had been entrusted to more traditional, established sectors—chiefly the Liberal Party—who were technocratic engineers and politicians with advanced degrees from abroad. To complicate matters further, Paraguay did not have an ambassador to Brazil for most of the duration of the negotiations. The previous government's ambassador retired in late 2008, and the Lugo government did not nominate a replacement until March 2012.

Lugo's coalition government and, specifically, the energy teams of his administration offered a "natural experiment" through which to observe the formation and effectiveness of two very different government logics. The new Itaipú negotiations had the power to reshape Paraguay in several ways. Not only were billions of dollars on the table, which could be used to alter the country's trajectory, shifting it toward a socialism akin to Venezuela's or Bolivia's, or any other number of options. But, also, Paraguay's relationship with Brazil was again in question. A new level of antagonism/distance or cooperation/integration with the smaller country's most powerful neighbor and trading partner would change its economic landscape, reverberating throughout the region. And so, the path of internationalism taken would have effects on economic and political structures inside Paraguay chiefly because of how decisions made there would have regional impact.

The ecological dimensions that accompanied and justified the arguments are particularly important to note: the changing meaning of water in a time of anthropogenic climate change; the changing value of reliable, non-carbon-emitting, domestic or at least regionally sourced energy in sustaining economic growth. As we look to how Paraguay renegotiated its relationship with Brazil in Itaipú, we will see that, though personal ambition and party loyalty had a role to play in creating heterogeneity in Paraguayan negotiating tactics, underlying the personal were ideological and, in fact, epistemological differences about the state. In the Paraguayan leftward turn, two views of the state struggled for dominance. Although Canese's team served as the original impetus for negotiations and the public face of the Lugo administration, and ultimately was the group that celebrated the binational Joint Declaration that resolved the conflict in July 2009, the document signed by Presidents Lugo and Lula was also incubated over months by the Itaipú technocratic team led by Balmelli. And so, to simplify the situation: one entity (the Hydroelectric Commission, headed by Canese) had a de jure while the other (Itaipú's directorate, headed by Balmelli) a de facto ability to deal with Brazil. Both Balmelli and Canese are important Paraguayan intellectuals, theorizing and historicizing Paraguayan politics. This chapter focuses on the two teams, their negotiating goals and tactics, their pressure points and publics, in order to examine how two parts of the left turn envisioned state power and just where energy fit into it.

The first decade of the 2000s saw an unheralded wave of democratically elected left-of-center governments take power throughout Latin America. Striking among many of them was the prominent role of natural resource management—and especially energy exports—in both rhetoric and the governing techniques of the rulers: Hugo Chávez (1999–2013) and petroleum in Venezuela; Evo Morales (2006–present) and natural gas in Bolivia; Rafael Correa (2007–17) and petroleum in Ecuador; Nestor Kirchner (2003–7), who originated from the hydrocarbon-rich Santa Cruz province in Argentina; Michele Bachelet (2006–10, 2014–18) and copper in Chile; Lula da Silva (2003–11) of the Workers Party in Brazil. And though Tabaré Vázquez (2005–10, 2015–present) of the Socialist Party in Uruguay did not explicitly campaign on a natural resource pledge, water and energy soon dominated Uruguayan international relations. As Paraguay was a latecomer to the leftward turn, the question asked in that country was whether Lugo would favor a Venezuelan-Chávez model antagonistic to foreign business interests or a more conciliatory Brazilian-Lula approach.

In spite of the vast differences between these governments and the specific national contexts that gave rise to them, the move to a new left came out of deep dissatisfaction with the Washington Consensus neoliberal economic order. Consequently, regimes throughout the hemisphere launched political-economic experiments in search of new economic models that would spur national growth and social development. Hydroelectric sovereignty and the renegotiations around Itaipú exemplified the pragmatic struggles to find a mechanism that avoided the vicissitudes of state-led socialism while still bringing social justice, and that embraced market-based opportunities while still protecting the national against globalization's predation. The important place of natural resources in all these governments hints at an "ecoterritorial" shift that will characterize governance in the twenty-first century as countries confront the impacts of anthropogenic climate change. Early on, in an important *Foreign Affairs* article on the pink tide sweeping across South America, Jorge Castañeda noted that there were two lefts, not one, which he described as either reformist and internationalist, stemming from the traditional left of the past, or nationalist and close-minded, another incarnation of Latin American populism.[2] Even before the leftward turn, Kurt Weyland found a deep affinity between neoliberalism and populism because they are both "political strateg[ies] with low levels of institutionalization" and because, through promising market reforms, charismatic leaders accrued legitimacy from the people.[3] That is, populism runs deep in Latin America, regardless of political affiliation. Natural resource management holds populist potential for the Right and for the Left, and not only in the Global South.

Arturo Escobar has characterized the left turns as crises of neoliberalism and modernity. He highlights the importance of indigenous identity politics in addition to the anti-neoliberal political economic projects as well as "development models that involve an ecological dimension" in the cases of Bolivia, Ecuador, and Venezuela.[4] With that in mind, he suggests that scholars look for two trends within the new politics: first, "alternative modernizations" that reject neoliberal economic models; second, "decolonial projects" that reject Euro-modernity values and practices in favor of communal, indigenous, and intercultural forms. Escobar builds his categories from Walter Mignolo, who offers a formulation different from the "right left" and "wrong left" descriptions of Castañeda, instead calling the political options the left, the right, and the decolonial.[5] Yet even the most anti-imperial sectors in Paraguay left hydroextractivism itself unchallenged. No one suggested dismantling the hydroelectric dams, and they eagerly spoke of Paraguay's untapped natural gas resources to the north. One question that remains to be answered is who may make claims on a state for resources (i.e., what constitutes citizenship and who is included and how in the national community). In the Itaipú case, hydrostate effects are not contained within nation-state dimensions but take on binational dimensions (regional) and thus require a negotiation of integration.

If the uncertainty about Lugo was whether his government would be more like Venezuela or Brazil, the answer was: both. Or perhaps, neither. Political shifts on the scale of the Latin American leftward turns provide opportunities for new state forms to emerge; out of the crisis of modernity, new political-economic formations are possible. In Paraguay, the ecological dynamics of development models were prominent and contested, a suggestion of water conflicts to come. Canese's involvement in Itaipú was expected—he was a recognized engineering expert with moral authority forged by persecution from the country's corrupt elite. But, to the surprise of many, Lugo named Carlos Mateo Balmelli executive director of Itaipú. The appointment immediately met with disapproval: former Liberal Party senator Domingo Laíno declared that Canese should have been nominated; Lugo's candidate for foreign minister stepped down in protest. Though Balmelli lacked expertise in hydroelectricity or Canese's engineering degree, he did possess a foreign PhD in political science. He wrote his doctoral thesis for the Johannes Gutenberg University Mainz (Germany) on the subject of Latin American constitutions and democracy and had been selected, after the fall of Stroessner, as one of the delegates to the 1992 assembly responsible for drafting the first postdictatorship Paraguayan constitution. Balmelli brought with him a technocratic team of engineering and finance specialists, including affiliates of the Colorado Party, who had decades of experience in Paraguay's energy sector and in government.

The partisan rules of Paraguayan politics, which require that resources and public offices be distributed according to a clientelistic logic, also set up competition within the executive when patron-client relationships fell across political parties. For all its inefficiencies, the Colorado Party had allowed for a kind of unity. Appointing a Canese-led Hydroelectric Commission under Paraguay's Foreign Ministry and a Balmelli-led Executive Directorate in Itaipú resulted in vastly different international strategies that came from divergent analyses of the situation with Brazil, of what kinds of political pressure worked, and ultimately of what the ideal relationship of Itaipú to the Paraguayan state and nation ought to be. And so, rather than merely bringing the large body of state debate literature[6] to bear on the choices of Canese and Balmelli's teams, in this chapter, I draw two practical and theorized understandings of state power from two of Paraguay's leading political thinkers.[7] Both were extremely intelligent, well-read, and conscious of the historical importance of the changes they were proposing as they used Itaipú and integration with Brazil to implement theories of statecraft. Importantly, they both are members of the New Democrat class Hetherington describes as a key liberalizing group in the democratic transition post-Stroessner.[8]

This chapter counterposes the twin campaigns by the Ministry of Foreign Relations and by the Itaipú technocratic team, highlighting the intellectual groundwork as members of both groups analyzed their own tactics. Both teams sought to change the public imagination regarding state and power, how government operated, what electricity was for and, ultimately, they argued for a new kind of regional integration. Both groups comprised a number of pragmatic scholars, some of which we will see in the ethnographic vignettes below as they intentionally invoked political philosophical literature to describe how they thought state power worked and to narrate their own actions. The political process by which the Paraguayan government defined the goals for negotiation resulted in the Six Points Memorandum, a canonization that simultaneously entrenched silences. After the Six Points were drafted in mid-2008, both teams commenced their negotiation strategies. In these two negotiation campaigns—publicly performed, planned and executed in private—competing political ideologies and professional expertise resulted in strikingly divergent strategies. Moreover, the very goals of their negotiations differed, as can be best seen in the language of the desired outcomes. But neither strategy fit neatly into the categories of "alternative modernizations" or "decolonial projects"; instead, illustrating the diversity of New Democrat thought, they forged distinct balances between market incentives, social development, and state involvement.

The Commission on Hydroelectric Binational Entities

THE SIX POINTS MEMO: DEFINING HYDROELECTRIC SOVEREIGNTY

The Six Points Memorandum presented by Lugo's government to Lula's in August 2008 produced an official account of hydroelectric sovereignty for the first time in Paraguay. Canese and members of the Paraguayan Ministry of Foreign Relations, supported by the Commission on Hydroelectric Binational Entities (CEBH), began meeting with Brazilian Foreign Ministry (Itamaraty) negotiators once a month in August. Although Itaipú's Brazilian executive director Jorge Samek attended these meetings, Carlos Mateo

Balmelli was conspicuously absent. And so, by the time I was able to speak with Canese, the two delegations had met several times. In an administration later plagued with accusations of slow decision making, the alacrity and determination with which renegotiations over Itaipú began was all the more noteworthy. A summary of the Six Points was immediately published in the daily papers, and they became an official metric for the success of the negotiations.

Though Ricardo Canese used slightly different terms in our November interview, the Six Points are:

1) Unrestricted availability (*Libre disponibilidad*). The full freedom for Paraguay to access and use its half of the energy as it wishes—the ability to choose the market.
2) Fair price (*Precio justo*). A fair (or just) price for the energy.
3) Revision of the debt (*Revisión de la deuda*). Reviewing (and revising) the debt incurred during the construction of the dam and renouncing any spurious or illegitimate debt (see chapter 4 for more detail).
4) Coadministration (*Cogestión*). Equal administration of the dam, alternation of top positions.
5) Control (*Contraloría*). Allowing the Paraguayan comptroller access to audit the books of Itaipú.
6) Pending projects (*Obras faltantes*). Completion of pending projects that were stipulated during the negotiations of the 1960s and 1970s.

The pithy list of simple phrases stands in contrast to most government communiqués in Paraguay, where excessive ceremoniality is the norm. Extremely formal address, showcased and subtly mocked in Augusto Roa Bastos's masterwork of Paraguayan fiction *Yo, El Supremo*, characterizes the public expression of governance and authority in Paraguay.[9] It weaves throughout bureaucratic ritual and the texts of secret police files. In Roa Bastos's novel, personal servant Patiño switches between florid honorifics, from "Your Mercy" (*Su Merced*) to "Excellency" (*Excelencia*) to "YerExcellency" (*Vuecencia*) to "Most Excellent Sir" (*Excelentísimo Señor*) to address "The Supreme" (*El Supremo*), Paraguay's first dictator Gaspar de Francia (1814–44).[10] The indirect address and passive voice of these early conventions lingers in Paraguay. And even when asking politicians and upper-level bureaucrats to meet as part of my research project, the key turn of phrase that elicited positive responses was to "beg an audience" (*pedir una audiencia*) of someone rather than "asking for the opportunity" to meet.

The difference could not have been more striking. And it was brilliant marketing. Each of the six issues condensed a series of complaints, and this summation left space for flexible interpretation of both the past and the future (hence, disagreements about whether or not the points had been accomplished). By listing them in order of decreasing importance and by placing the most contentious issue at the beginning, the Paraguayan negotiators clearly marked the limits of their aims. Breaking their claim into six parts would allow Brazil to gradually compromise without being seen as surrendering too much too quickly and would establish a precedent for how to yield to Paraguay. This also allowed Paraguayan politicians to assert that progress had been made even as the process stalled (perhaps indefinitely) on the edge of the core issue.

What Paraguay's government ultimately wanted—*libre disponibilidad*, unrestricted availability (Point 1)—was the ability to sell Paraguay's electricity at the price it chose (Point 2) to whatever market it chose. This was currently impossible, both legally and physically. Because it only had four 220 kilovolt lines at the time, Paraguay lacked the necessary high-tension lines to acquire and transmit its half of the electricity within the country. The infrastructural obstacles (Point 6), however, were minimal compared to the political and diplomatic obstacles to unrestricted availability (Point 1) and a fair price (Point 2). The 1973 treaty, which would require revision in 2023, held that only the two countries who shared Itaipú could benefit from its energy, and so, by contract, whatever electricity Paraguay did not consume had to be sold exclusively to Brazil. In the interview that opened the chapter, Canese himself referred to "unrestricted availability" as "[hydro]electric sovereignty," the ability for Paraguay to decide how the nation's natural resources should be used. For the Hydroelectric Commission, for Lugo's progressive allies, and in the framing of Itaipú by the national media, Paraguay's hydroelectric sovereignty was violated by the treaty restrictions.

But more was at stake than just the first two points. A binational entity presents a curious problem in nation-state sovereignty in its relationship to the two states to which it supposedly is subordinate. For its entire existence prior to Lugo's presidency, Itaipú Paraguay was in the hands of the Colorado Party. Loyal party members got multimillion-dollar construction contracts and appointments to high-earning positions. Critics complained that Itaipú also directly financed the construction of private homes, excursions for the political elite, and the political campaigns of Colorado candidates. All this was carefully documented on official letterhead with multiple stamps and seals and filed away in Itaipú; a paper trail clearly led from the coffers of

Itaipú to Colorado Party campaign expenses. In order to avoid scrutiny, it was argued that, since Itaipú was binational, it lay outside the jurisdiction of the national comptroller and the district attorney. For decades, the expense books of Itaipú were left untouched and inaccessible even though newspapers reported that high-level employees would suddenly appear with new cars that cost more than their annual salary. Balmelli had once explained to me that "the collusive financial mismanagement between Itaipú and the government, mediated through the Colorado Party, was not a desperate secret that all involved members tried to hide; it was the glue that brought different actors together and the motive for acting together in governance." And this is what Point Five (control) was intended to check. If Paraguayan government accountants were regularly denied access to Itaipú's daily expenses, one might imagine the initial reticence of Eletrobras to allow Itaipú's construction debt to be examined (Point 3) and disowned if found spurious.

Beyond the political nature of the appointments to the Paraguayan board, there is a subtle difference in titles that reflects a substantive difference in status: the word "executive." The most powerful directorships—the executive financial director and the executive technical director—had always been Brazilian; the positions of executive legal and administrative directors always Paraguayan, as was the executive director of coordination. The disparity between the two halves of the Executive Directorate meant that Brazil's directors initiated the most important decisions. It also went in tandem with Brazil's control of the finances behind the construction of Itaipú. Through demanding coadministration (Point 4), Paraguay sought to rectify this imbalance.

From their first announcement, the Six Points became the framing device when Canese and the Hydroelectric Commission team explained the state of the negotiations with Brazil to the public.[11] Concretizing the terms of debate in this way was a strategic move, a state-initiated code switching to shift the issue of Itaipú from its past life as a cause célèbre of the opposition to the central diplomatic project of a serious government. The register difference signaled, both within Paraguay and to an external audience, an appropriation and legitimization of hitherto opposition critiques by sanitizing them. That—in spite of its loudly protested objections to the Six Points, Brazil's government would immediately agree to treat with Paraguay's on issues that could conceivably end in billions of dollars of lost revenue the very first time such a request was made of Brazil—arose in part from Paraguay's new reputation as a maturing multiparty democracy.

As the Six Points consolidated an official version of hydroelectric sovereignty, the act of stamping a story with an official seal rendered other stories apocryphal. Hydroelectric sovereignty here was a question of balance sheets, price points, massive feats of engineering, and division of labor. Honing the scope of the negotiations by formulating sovereignty as a matter of finance, physics, and corporate hierarchies—in sync with the most accepted authorities of the day—shunted other issues, relegating them to the margins. Questions about the environmental impact caused by the flooding of the Paraná or about the human impact caused by the dislocation of communities that formerly lived on the riverbanks were thus not part of international negotiations. These notable silences indicate compromise that may make it all the more difficult to link hydroelectric sovereignty to issues of the environment or human rights.

THE CEBH IN ACTION

Brazil's negotiators initially refused to budge on the first three points, which both countries saw as the heart of the matter. And so the rounds of negotiation continued between the Hydroelectric Commission and the Brazilian foreign ministry's team as Paraguay's negotiators rejected the counteroffers extended as palliatives: doubling the price Paraguay received for energy ceded to Eletrobras, and the creation of a fund to underwrite development projects in Paraguay. As we saw in chapter 2, the Brazilian negotiators defended their intractability with two arguments. First they argued that, whereas the Paraguayans were trying to violate a signed treaty by renegotiating it, the Brazilians were in good faith upholding the treaty and were thus the victims. Second, they asserted that because Brazil supplied the collateral for all the construction loans, it alone assumed the risk should the dam project fail, and in this way Paraguay was trying to get something for nothing. And so the central riddle Paraguay's negotiators had to solve was how to change Brazil's posture, what combination of pressures and inducements would suffice.

Although highly stylized ceremony characterized initial interactions with state officials in Paraguay, it was counterbalanced by sometimes startling spontaneity. On the December morning I phoned Cayetano Pujols, a Ministry of Foreign Relations representative who was part of Canese's team, I was in a taxi to his office in less than twenty minutes.

As soon as I arrived, I handed him my card, and, while he returned the gesture, I said, "Thank you for meeting with me so quickly. I was surprised."

"Only in Paraguay," he replied, with a hint of dry humor, "can you call and have a meeting with a government functionary fifteen minutes later."

Then he leaned back in his chair and said, "Well. You don't seem to be from Brazil."

"No, I'm obviously from the States." I said the last word in English, with a tell-tale accent.

"No, I mean that you don't appear to be on the Brazilian side of this. So, you're an anthropologist."

"Yes, but what I study isn't what most anthropologists [in Paraguay] work on. Most anthropologists here work on indigenous people and rural subjects. I work on the urban and the state. . . . I study the state as it is, in practice, not just in theory."

"Michel Foucault," Pujols said at this point, "writes that power is not located in a place; the state is a strategy. Have you read Foucault?"

"Yes," I replied.

"Well, and so, taking the control of the government, as happened with Lugo, doesn't solve the situation."

"Right, that's why occupying a building leaves you with . . . well . . . the building."

"It's just a symbol," he concurred.

Seemingly satisfied with my credentials, Pujols turned to the subject of Itaipú, saying, "There are other Paraguayan voices that can be made content by smaller concessions, some money. But we are firm. We are not just talking about money, but modifying the relationship Paraguay has with Brazil." That is, though rent might have been the primary accumulation strategy of the Paraguayan government, the larger priority was to change the relationship of Brazil to Paraguay in the minds of Brazilians and Paraguayans.

Then Pujols explained the strategy of the Hydroelectric Commission: "Canese is trying to mobilize social groups and the universities in favor of renegotiation. We are in conversations with Brazilian social movements, too. To discuss this in an international plane, you need to have adepts in other places." Foucault also reminds us that state power is capillarized, located not just in halls of power, but in the everyday practices of a population living under rule as well as in spectacular moments of protest and uprising. As a diplomat, Pujols had access to the inner workings of the negotiations, not just the public strategy. "Brazil is a power," he said. "Itamaraty [Brazil's Foreign Ministry] has a school of diplomacy where people are trained. Nevertheless, we've left them unable to speak many times because our arguments are reasonable. They invoke international law haphazardly, inconsistently. And international law favors Paraguay."

To help illustrate what these encounters were like, Pujols told a story: "I remember at a meeting earlier what my Brazilian counterpart said to me. . . . And I recall what he said to me about Point 1: 'It is unthinkable' "—as Pujols relayed the interaction, he placed a strong emphasis on the last word—" 'that Brazil will become flexible. This was a strategy for both countries to get electricity; no one even gave a thought about unrestricted availability or sovereignty under the document signed under two dictatorships.' I won't tell you what I said to *him* in response." In other meetings with Hydroelectric Commission strategists, some gave me copies of books or white papers or outreach materials, others laid out imperial chronologies from colonial Spain to the present. From citing Foucault in order to make sense of the mystifying power of the state to placing international law in an interpretive field based on the history of imperialism to deliberately reconceptualizing sovereignty, members of the Hydroelectric Commission did not merely happen on a strategy. Rather, as organic intellectuals with explicitly articulated counterhegemonic commitments, they applied Marxist theory and historical analysis in order to gain greater power for the working classes and campesinos.

Five months into the negotiations, the Brazilian Foreign Ministry made a formal proposal intended to outline the contours of compromise. On January 26, 2009, the Brazilian counteroffer was communicated to the Paraguayan Ministry of Foreign Relations. It contained three main points: doubling the amount received for compensation, creating a $100 million development fund (seeded by BNDES, the Brazilian development agency), and $500 million in financing for the construction of electricity infrastructure. None of these three propositions addressed the Six Points Memo, the questions of hydroelectric sovereignty, or unrestricted availability, or anything about market prices. Indeed, the offers strayed far from the central Paraguayan concerns and may have come from previous experiences where the promise of loans or construction assistance (which could be distributed along patron-client networks in Paraguay) served to placate the Paraguayan political elites, reminiscent of Brazilian inducements around the Itaipú fifty hertz/sixty hertz controversy of the 1970s recorded in chapter 1. The Lugo government rejected the offer.

To demonstrate the importance of communicating to the Paraguayan people, the Hydroelectric Commission dedicated a subcommittee to public relations, balancing the work that Pujols and others were doing in closed-door negotiations. The communications team embarked on a series of presentations and rallies throughout the country, visiting schools, using a newsprint comic rendering Brazil's bullying and the Six Points as a pedagogical instrument. The communications team's work was doubly necessary because the last meeting of the two delegations in late January 2009 ended abruptly as a result of mismatched expectations between the two. The foreign ministry teams only formally met again around the signing of the Joint Declaration in July 2009. Instead, Fernando Lugo himself, in a state visit to Brazil at the end of April, personally asked Lula da Silva to see to the matter. While the language of confrontation and imperialism and battle might serve in a Paraguayan public, in international diplomacy, it did not play as well. Even as the high-level meetings between the Foreign Ministry negotiators ground to a halt, the message of recovering sovereignty from an imperialist power and the threat of unilaterally renouncing spurious debt and a treaty signed under duress continued to spread within Paraguay.[12]

On March 26, the Hydroelectric Commission communications team helped organize an international rally in the eastern department of Alto Paraná, where Itaipú is located, an example of the mobilization of social groups and university students described by Cayetano Pujols. News of the rally spread through public and personal channels, brief articles in the press, and word-of-mouth to social movement leaders, announcing support for "popular integration with energy and food sovereignty, constructing the Great Homeland"[13] to try to pressure for renegotiation of the Itaipú Treaty. Campesino, indigenous, worker, and progressive groups from throughout Paraguay (with a strong showing from local communities) gathered in the plaza across Ciudad del Este's municipality before marching down the international highway toward the Friendship Bridge (see figure 3.1). As they carried banners—some printed by design shops, others written by hand—down the thoroughfare usually crowded with buses, vans, taxis, and moto-taxis shuttling day tourists between Paraguay and Brazil, vendors sipped ice-cold *tereré*, and the police looked on. The Friendship Bridge, spanning the Paraná River just 8.7 miles south of Itaipú, was closed to all vehicular traffic as a crowd of about one thousand from Paraguay and a few hundred from Brazil gathered, straddling the line between both countries. *Chipa* vendors, carefully balancing brimming baskets of bread on their heads, wandered through the crowd, as did peddlers hawking pins with Brazil and Paraguayan flags and a man offering straw hats brimmed with ribbons pro-

FIGURE 3.1. Rally for hydroelectric sovereignty at the Friendship Bridge in Ciudad del Este. *Source*: Author's photograph, March 26, 2009.

claiming "El pueblo unido jamás será vencido" (the people, united, will never be defeated).

A Portuguese-language flier from Vía Campesina explained "Why we are mobilized": "We assert that the people of Brazil, Paraguay, and other Latin American countries are one single people. . . . Integration, yes! But of the people, not of the markets. . . . The dream of the Great Homeland of Jose Martí, Simon Bolivar, Che Guevara, Dr. Pedro Casaldáliga." The yellow/green/blue and the red/white/blue of the national flags intermingled with the red flags of Brazil's Landless Workers Movement (MST) and the green of the Paraguayan Campesino Movement (MCP) as the rally formally began around 9:30 a.m. with the blaring of Brazil's and Paraguay's national anthems from speakers mounted on a truck. Brief speeches—either in Spanish, Portuguese, or the regional blend of the two, Portuñol—delivered by male representatives from both countries declared the "Great Homeland" (Patria Grande) and solidarity between all peoples of South America, denounced neoliberal empire, and clamored for "complete reform" (*reforma integral*), including agrarian reform and the recovery of hydroelectric sovereignty.

"The people of this continent Latin America are constructing another Latin America in the interests of the people," proclaimed a Brazilian speaker from the MST. "There is only one struggle. . . . The struggle for sovereignty over energy resources, over water resources, is a common struggle." And then he cried, "Long live the integration of Latin America!" and the people cheered.

"First, brothers and sisters, I want to celebrate the support for the Paraguayan cause given by the presence here of the Brazilian movements and of Vía Campesina as well as those from Argentina," began the next speaker, a Paraguayan. "A strong applause for them!" The crowd complied vigorously. "We are here to demand that the governments of Brazil and Paraguay renegotiate the treaties of Itaipú and Yacyretá and to recover energy sovereignty. . . . We denounce the oligarchy . . . of industry in Brazil and the oligarchy in Paraguay. . . . And the oligarchy installed in Itamaraty."

The last of the speeches, by a Brazilian leader of the Movement of Dam-Affected People (MAB, Movimiento dos Atingidos por Barragems), was followed by a symbolic exchange of national flags, hugs "in solidarity," and then good-byes "until next time." Although MAB formally began in 1991, Brazilian indigenous and peasant communities affected by dam construction had begun mobilizing on themes of environmental destruction, popular sovereignty, and economic justice against the disruption caused by hydroelectric projects across the country in the 1970s, including Itaipú. The MAB has grown over the years and now hosts international assemblies and even helped launch an annual International Day of Action against Dams (March 14). In spite of MAB's global reach, their brand of social justice did not receive as much airtime in Paraguay outside of the March 2009 rally.

By 10:15, the protesters were clearing the bridge and winding up the international highway before regrouping at the plaza; the business of Ciudad del Este sprang back to normal. A smaller Paraguayan crowd at the Plaza de Paz continued the protest with leaders (including a woman) taking the mic in Guaraní, to continue the call for agrarian reform in addition to hydroelectric sovereignty. The day's events targeted the center of commercial activity in Paraguay and the visible symbol linking Brazil to Paraguay. Though it disrupted traffic for a few hours, the quick clearing of the bridge mitigated the financial impact; most day shoppers enter Paraguay by 7 a.m. (PT)/8 a.m. (BT), and the massive cargo trucks carrying agro-exports begin lining up around 4 p.m. While billed as a binational march, the majority of the protesters returned to the Paraguayan side of the bridge. The Paraguayan press front-paged photographs of the protestors: human bodies filling shots of the bridge or roads; in Brazil, nothing on the covers of the major dailies.

The schoolhouse meetings, public presentations, and Ciudad del Este mobilization of social movements in Paraguay and in Brazil, the tactic Pujols had explained long before, used a similar idiom. There were strong critiques of the capitalist order, affirmations of the transnational unity of "the people," and a repudiation of the current system. The language on the Friendship Bridge was one of "popular struggle" and "oligarchy." And the core aims were those published by Canese years before and augmented at the working group meetings—a "fair, market price" for energy, the renunciation of "spurious debt," and a renegotiation of the Itaipú Treaty, all as a sign of popular sovereignty.[14] Rhetorically, the speeches addressed Presidents Lula and Lugo. However, given that Brazil's media ignored? did not notice? the protest far from its economic, cultural, or political centers, the (intended?) audience was solidly Paraguayan.

The Ministry of Foreign Relations' Commission on Binational Hydroelectric Entities used the Six Points as a pedagogic tool to raise awareness about Itaipú in Paraguay, elevating it to a national cause that would draw citizens to the streets for protest. In the language of the Hydroelectric Commission, it seemed that the chief meetings where Paraguay achieved its sovereignty were those between the Brazilian and Paraguayan Foreign Ministries negotiating teams, as well as through the public moments of united Paraguayan protest. The international goal of public protests made a moral appeal to popular sectors in Brazil so that they would join Paraguay in clamoring against the mistreatment of Paraguay in the Itaipú Treaty. On a national scale, rallies and teach-ins built consensus while establishing the authority of the leftist activists in the Hydroelectric Commission. In rhetoric and action, the commission acted as if the pressure point in both countries was the people (*pueblo* or *povo*), chiefly working-class and campesino sectors who had been oppressed by oligarchic capitalist interests that also controlled the state apparatus. And should the capitalist-controlled Brazilian state not accede, punitive measures would escalate.

Itaipú's Executive Directorate

Itaipú executive director Balmelli had the opportunity to make his case just a few weeks after our first interview. For much of mid-November 2008, the activity in the antechamber to his office, where three assistants and several bodyguards worked, was more frenetic than usual. The head of communications, whose office was on another floor, spent the days racing between the two floors, shuffling papers, and making telephone calls. The crisis of the week was the discovery of maintenance men illegally burning boxes of Itaipú files per the orders of they-wouldn't-say-who on a sleepy Sunday afternoon, purportedly destroying evidence of fiscal malfeasance from the previous administration.

But though the sabotage against Paraguay's ability to hold its leaders to account was urgent, there was something even more important on the executive director's agenda. He was preparing for an international trip (other than Brazil), his first in his official capacity. The plan was to present a legal argument, which Balmelli described as "logical" and "rational" (as opposed to arguments based on national pride or denunciations of imperialism) in an international platform, thus elevating the issue beyond its previous scope—a calculated move that would frustrate Brazil's attempts to contain the debate. The parliament of Mercosur (Parlasur) had scheduled a public discussion between the two executive directors of Itaipú in Montevideo, Uruguay, out of a concern to understand the escalating diplomatic tensions between two of its founding member countries. Mercosur delegates from Argentina and Uruguay joined representatives from Brazil and Paraguay on November 29, 2008, to hear paired forty-minute expositions, first from Jorge Samek and then from Carlos Mateo Balmelli.

The Brazilian executive director's presentation centered on the finances of the dam. With a twenty-slide PowerPoint presentation, Samek challenged the claim that Paraguay only received $2.81 per megawatt hour in compensation for energy ceded to Brazil by explaining the components of the energy tariff that Paraguay also received as part of payment but that went to pay royalties, salaries, the construction debt, and social development projects (see chapter 2). Chart after table showed changing interest rates and line graphs. To definitively demonstrate the role of Itaipú as a tool of development, Samek highlighted the amount of money both governments had received from the dam since operation in 1985: Paraguay's government had received $4.8 billion and Brazil's $3.9 billion up to that point, in addition to income from energy sales by their utility companies. More than thirty-five minutes of the presentation focused exclusively on the numbers—energy production amounts, tariff components, interest rates, indicating that the argumentation Samek thought appropriate (and convincing?) was one of economics and statistics.

When Balmelli took the stage, he opened a different line of reasoning, one prepared in conjunction with his engineering and technical advisers.

"And so, after thirty and some odd years comes the doctrinal discussion of the famous *pacta sunt servanda*: treaties must be kept and applied. And the famous doctrinal clause: *rebus sic stantibus*," he said to the assembled international body. "Obviously, the rigid interpretation of the treaty is *pacta sunt servanda*; this does not merit interpretation. On the other hand, the clause *rebus sic stantibus* refers to the fact that there are new circumstances. And I believe, sirs, that there exist new circumstances to rethink the application of the treaty. In the first place, as we have seen, the value of electricity is different in an expanding economy. But, above all, the value of electricity is different from those of other kinds because it has characteristics that make it useful for any market. . . . And so, it is admissible because it does not contaminate; it is accessible because the water is there, available, and permanently generating."

Balmelli invoked the legal principle of *pacta sunt servanda* and the carefully used challenge *rebus sic stantibus* to explain his argument that the Itaipú Treaty should be enforced, but that the way it was enforced could be different from how it had been in the past, particularly given changes in regional politics and in the value of hydroelectricity.[15] Through this he argued for a new interpretation of the existing treaty: that Paraguay be allowed to commercialize Itaipú electricity on the Brazilian market via ANDE.

Several times throughout the presentation, he said, "Let us apply the treaty," contrasted with the Hydroelectric Commission's calls to renegotiate a "bloody" treaty that had been signed by two dictatorships. Based on this platform, Balmelli called for "energy integration" as part of a broader move toward integration not only in Mercosur, but in Latin America, as Lula himself had oft mentioned. And as Brazil stepped into a leadership role in the midst of the region, Balmelli argued for a relationship of "partners" (*socios*) not "satellites" (*satélites*). Nothing about land appropriated during the War of the Triple Alliance or renunciations of billion-dollar loan commitments or threats of going to The Hague was mentioned. Instead, Balmelli referenced the global energy crisis and the need to generate solutions by expanding "good neighborliness" even as Brazil rose as a global player, surrounded not by "peripheral satellite" countries, but by "business partners." In addition to Dependency Theory language, he invoked nineteenth-century Argentine Juan Alberdi's prediction that continental equilibrium would not be political-military, but rather an evening of economic, political, and social benefits from commerce and trade.[16]

The importance was not lost on his listeners. Senator Juan Domínguez from Uruguay said, "The solutions that we are able to find in regards to

Itaipú will serve for another series of problems that we have" and clarified later by explicitly referencing the Guaraní Aquifer and Bolivia's natural gas.[17] The words of Alberdi and questions of hegemony struck a Brazilian chord that very day. Brazilian parliamentarian Marisa Serrano (a member of the opposition to Lula's government) said, "When your Excellency spoke of a continental equilibrium . . . these are words with which I fully agree. . . . And I do not see in Brazil any idea of hegemony, quite the reverse."[18] Even for someone distant from Lula's policies, it was necessary to countervail any whiff of Brazilian unilateralism or imperialism.

The argument before Parlasur was between financial data and legal prospects, a dry discussion about future investments and minutiae of logical extrapolations from decades-old treaties. The language of applying the treaty, though unpopular within Paraguay, and many of the changes the technocratic team in Itaipú's Executive Directorate pushed for (selling via ANDE to the Brazilian market) eventually became the palatable solution for the presidents of Paraguay and Brazil when they signed the Joint Declaration of 2009. But the legal component was only part of Balmelli's argument. It linked to a question of the kind of regional integration and international relationships Brazil wanted to establish with its neighbors in its ascent as a rising global power: a question of unilateralism or multilateralism.

Balmelli had also been invited by the Casa América in Madrid to give a speech a few days after the Montevideo debate. The Madrid-based consortium serves as a forum for Ibero-American issues ranging from art and culture to politics and economics. His speech before business and political leaders from the Spanish-speaking world made the European media, and a high-quality video of Balmelli's talk was quickly posted online. "What is it that we are asking?" he asked. "The application of the treaty with equilibrium between the members and equity in the treatment of all the partners that form part of the treaty, in this case, Paraguay and Brazil."[19] After Spain, it was to Vienna that he would travel, again with his counterpart Jorge Samek, to cosign a Memorandum of Understanding (MOU) between Itaipú and the United Nations Industrial Development Organization (UNIDO). He saw these three different international stages as distinct opportunities to initiate a new tack for the struggle over Itaipú.

"Sovereignty," he said to me as he prepared for the trip, "must be exercised. That is what I am doing." According to Balmelli, it was vital for Paraguay to present itself as co-owner of Itaipú in front of the rest of the world by not allowing Brazil to be the only signatory of the MOU between Itaipú and UNIDO in Vienna. A recent advertisement for Eletrobras in the *Financial Times* that announced the company's inclusion in the New York Stock Exchange with the caption "Brazilian Energy, now in New York" underneath a photograph of Itaipú had elicited uproar in the Paraguayan press.[20] This was seen as yet another attempt by Brazil to present itself as sole proprietor of the dam and an insult to Paraguay's sovereignty. And to be sure that a Paraguayan audience saw all this, Balmelli was accompanied by an adviser on loan from the Ministry of Foreign Relations, who photographed and recorded the entire trip.

The public audience at Parlasur was more than just an opportunity to speak in front of regional politicians. The fractious history of international water politics itself had given rise to ideas of regional cooperation that led to Mercosur. The parliament of Mercosur, a newly formed body with unclear jurisdiction in a common market of questionable effectiveness, took uncertain toddling steps toward regional relevance by engaging on the issue traditionally at the heart of Southern Cone international dilemmas. Although Parlasur's decisions exercise no binding weight on the member countries and though the presentations and speeches in Montevideo garnered but passing attention in the Paraguayan press, these events laid legal and theoretical groundwork for the July 2009 Joint Declaration signed by Lugo and Lula. And this not merely because it was a gathering of regional political elite, but because the Southern Cone countries have experience working out international balances of power over water in both bilateral and multilateral contexts. What Parlasur was doing in calling the meeting was testing its strength.

"We can't crash against Brazil," Balmelli said in December 2008 once he returned to Paraguay after his trips abroad. We again sat in his corner office in Asunción. In the middle of our conversation, Diego Ferrer (the technology assistant in the antechamber) brought me a full-color photocopy of the article about Balmelli written in the Spanish daily *El País*.[21]

"Of course, there's an asymmetry between Brazil and Paraguay," he continued. "I would be stupid to not recognize this. . . . But, if you interact between two states, it is not the same as between two human beings. Some people think it is, but human behavior is not state behavior. This has been said since Thucydides wrote about the Peloponnesian War. The commission was emotional, but you have to be cool and relaxed." Like others in Itaipú's Executive Directorate, Balmelli shared a dismal assessment of the current state of negotiations between the Hydroelectric Commission and Brazil's

Foreign Ministry team. As he reflected on the outcomes of his international trips, he gave his opinion on the negotiating goals of the Hydroelectric Commission. "For Paraguay to refuse the debt—this would be the best scenario for Brazil to not pay Paraguay for its energy. . . . This is naive," he said.

"Is Paraguay looking to follow the model of Ecuador?" I asked, because I had already heard comparisons between Paraguay's negotiations with Brazil and the negotiations of other border countries with the South American giant.

"The situation is very different," Balmelli countered. "Paraguay is no Ecuador or Bolivia. Ecuador owes $200 million. Bolivia $6 billion, and Petrobras is in Bolivian territory. And it's not like the Panama Canal, which was in Panamanian territory, and the treaty was expiring. That was a question of respecting the treaty. This is different: $60 billion [i.e., the Itaipú debt] and a border." Whereas the Hydroelectric Commission saw precedents for Paraguay in other international negotiations over valuable resources, Balmelli saw uncharted waters.

And even in December 2008, he mentioned to me a robust explanation for how Itaipú ought to develop Paraguay. "I don't want more money," he said. "It is better for Brazil and for Paraguay for us to improve our economy via regional integration. What we need is to move Paraguay beyond viewing Itaipú as a source of rent. . . . Instead, the energy sector is what modernizes the economy. Because there is not enough public sector, the investments need to be from private firms. This is through energy integration, which is the concept of *seguridad energética* [energy security]." The idea here was that, through using more of its Itaipú electricity for industry and by having ANDE mature as an organization through competition on the Brazilian market, Paraguay's politics and economy would develop beyond its patrimonial and parochial past.

From November 2008 onward, in ordinary meetings with the Brazilian Itaipú Directorate or special presentations to international audiences, "article XIII," "enforcing the treaty," "business partner," "let us sell to the Brazilian market," and "integration" were the mantra of the new leaders in the Paraguayan Itaipú Directorate. Implicit in these encounters was more than the reliance on legal and market-oriented logics to win the day. The Itaipú team acted as if the power brokers whose opinions had to be swayed in order to effect change in Itaipú were Brazil's economic and political elite who could be convinced through legal and market-oriented logics and pressured through the international press.

In late January 2009, just before the formal meetings between the two foreign ministry delegations came to a standstill, Balmelli recounted to me a recent conversation he had had with Paraguayan president Lugo.

> And I said to Lugo, "Think about the scenario where you will treat. Remember, you can get something with Lula. If the conservatives win in Brazil, forget it. If the conservatives campaign and complain, 'Lula was weak with Correa, Morales [i.e., the presidents of Ecuador and Bolivia],' the candidate [i.e., Dilma Rousseff] that Lula is supporting will ask Lula not to touch Itaipú because 'We will look weak.'"

This real politik assessment of Brazilian politics seems to have been alarmingly prescient as conservative sectors in the government—skeptical of the concessions Lula eventually made—impeached Dilma Rousseff in 2016 on corruption accusations.

Our conversation was interrupted, as it often was, by a phone call. Not only did Balmelli's office have two landlines; he also had three personal cell phones with different numbers. Switching into Spanish as he spoke with his caller, he explained why he disagreed with the Hydroelectric Commission's focus on demanding hydroelectric sovereignty. "Sovereignty?" Balmelli said. "If there's a market who will pay the market price, who cares who is the buyer? This will allow me to develop my country. . . . Paraguay voted for change. . . . We have to push for unrestricted availability, to sell in the Brazilian market. And for a fair price. This is what to focus on." For all the critique of the Hydroelectric Commission's argument, the Itaipú technocratic team readily used the phrasing of Points 1 and 2 of the Six Points Memo, indicating the traction that text attained.

Once he finished his call, he continued his comparison between the two negotiating strategies. "We advocate a more rational line of argumentation, whose objective is that Paraguay attain more benefits. The line in the newspapers is more radical; it is one of demands. The arguments in the papers and in the negotiating team lack technical bases."

"What do you mean?" I asked. "Give me an example."

"For example, the myth that Brazil pays $2.71 and then sells it for $300," Balmelli answered, referring to the price differential for Itaipú megawatt hours sold in Brazil. "And with *libre disponibilidad* [unrestricted availability], they talk of Chile. But the high-tension lines do not exist, and the energy lines will have to go through Argentina." By this he meant that the physical and legal infrastructure necessary for a transnational electricity sale did not exist, making the aim of exporting to a non-Brazilian market a practical impossibility. One week later, the formal meetings between the two foreign ministry delegations paused, but the Paraguayan Itaipú technocratic team continued to meet with their counterparts outside of the public eye.

In March 2008, the Itaipú Directorate negotiations took a more serious turn and expanded to include conversations with the Brazilian ambassador to Paraguay and other members of the Brazilian Foreign Ministry. (Around this time, the Hydroelectric Commission was putting final touches on the large international rally planned for Ciudad del Este.) "Tomorrow I meet with Samuel Pinheiro Guimarães," Balmelli said to me. "He is my old friend, for twenty years. He was director of the think tank in Itamaraty." Samuel Pinheiro Guimarães, the secretary-general of the Brazilian Foreign Ministry, was a lifelong Brazilian diplomat, serving under military and civilian governments for decades, and was known as an advocate for regional integration and, in particular, Mercosur. He was in town as an invited guest of the Paraguayan Foreign Ministry in anticipation of a forthcoming state visit of Lugo to Brazil. Balmelli planned to take advantage of the trip to meet with Guimarães's team informally.

The day after Balmelli's meeting with the secretary-general, I was back in the executive director's office. "Brazil stayed quiet," Balmelli said. "They listened, but had no [op]position. I told them about the capacity to sell energy from Itaipú by ANDE in Brazil. . . . I said, 'There is the possibility to sell in the other market. ANDE and Eletrobras can hire another enterprise to do the work on behalf of ANDE or Eletrobras. The treaty allows this. The only one requirement is that they be a Paraguayan or Brazilian company. If I'm asking for something crazy, let me know. . . . I don't want to take anything from Brazil that doesn't belong to Paraguay.' "

At that point, we were interrupted, and while Balmelli spoke to his caller, I flipped through a folder containing the presentation the Itaipú Executive Directorate had made. The opening page contained a list of a dozen individual proposals that referred to line graphs and electricity prices and the construction of new energy transmission infrastructure. Notably, the line items did not include references to hydroelectric sovereignty or the Six Points. Instead, the list was technical, explicitly mentioning articles from the Itaipú Treaty and numerical figures with precise decimal points. It centered on a plan to transform ANDE into an international energy company on the Brazilian market, selling electricity from Itaipú and then from other to-be-built dams. It also included a proposal to increase the price of compensation by 500 percent with the argument, "The influence of increasing the multiplication factor used for determining the compensation for energy ceded from 5.1 (as it is now) to 25.0 or 26.5 would impact the Brazilian consumer with an increase of 1.6 percent and 1.7 percent, respectively. But the effect on Paraguay is an additional $500 million/year" (field notes, March 19, 2009). Though I did not know it at the time, I was looking at an early draft of the Joint Declaration, a document that called for a higher price for ceded energy in the short term and a long-term plan to develop Paraguay by increasing domestic energy consumption and industrialization and to finance that through direct electricity sales in Brazil via ANDE.

The prospect of raising the price of energy in Brazil merely to transfer more revenue to the Paraguayan government was met with resistance. "I told Samuel, 'In the tariff to the consumer you have taxes, you have things that have nothing to do with energy,'" said Balmelli to me. He then got another phone call and related (in Spanish) the same to his unnamed interlocutor, adding, "They raise the price of electricity to double just by taxes—this is a key part of our argument." The moment his call finished, he rang engineering adviser Guillermo Velázquez to ask, "What is the exact amount of taxes in the tariff?"

Then he switched back to English.

"Brazil and Paraguay are coproprietors of the water from the river. How do we define how much Paraguay and Brazil get for renting water so that Itaipú makes energy? What criteria is fair?" he asked and then answered his own question: "Not the financial cost [of construction], but the bill for making energy." Balmelli moved the argument away from the cost of construction and thereby the debt (spurious or otherwise) to, instead, the value of energy, which had changed because of relative fossil fuel limitations and the expansion of the Brazilian and Paraguayan economies.

And, though I had heard it before, he reiterated his position vis-à-vis international law, saying, "Paraguay can sue Itaipú, not Brazil, because it hasn't adjusted the price. There are two legal principles: *pacta sunt servanda*—you have to respect, apply the treaty. *Rebus sic stantibus*—which means new circumstances bring to debate a new point of view to interpret, then this is a reason to change agreements. Paraguay uses *rebus*, but Brazil uses *pacta*. And I, this Paraguayan, use *pacta*." He continued as if in a fictional conversation with a Brazilian audience, "I say, 'Let me sell to Brazil. The Argentine market is not viable, if I sell, they won't pay. And we lack the *redes*, the network. I share the same position as you—*pacta*, I never say renegotiate the treaty.'" Just as the Itaipú technocratic team began a serious lobby for its proposal, politics in Paraguay descended into the chaotic. As a sign of the political instability, in early May 2009, the Paraguayan executive directors of Itaipú and Yacyretá were interpellated before congress, required to give account for allegations of questionable behavior. Balmelli passed his interpellation with strong support from most political sectors; the executive director of Yacyretá, a member of Tekojoja, on the other hand, received tepid support if not outright hostility. But as months passed, the behind-the-scenes conversations with Brazilian Foreign Ministry representatives continued.

On May 20, 2009, as Balmelli discussed the ramifications of the interpellation and the growing rift between the leftists and the Liberal Party in the Lugo coalition, one of his many phones rang. "Eduardo, my friend, how are you?" said Balmelli in Portuguese to the Brazilian ambassador to Paraguay. He then invited Eduardo dos Santos to dinner, interspersing some Portuguese terms with Spanish (using both "onde" and "donde" for "where," "charlar" and "falar" for "chat"). "The conversation with Samek went very well. The solution to Itaipú will be within Itaipú." Balmelli, Velázquez, and others in the Itaipú technocratic team cultivated personal relationships with Brazilian colleagues in order to build support for the Paraguayan proposal, acting as though, in order for the Brazilian government to risk the political capital implied in signing an agreement that could result in immediate price increases in Brazil, the industrial sector's buy-in might soften the blow. As the one-year deadline of the new Itaipú negotiations approached in late June 2009, the mood within the Itaipú technocrat team grew increasingly optimistic about their prospects.

"I think we will attain something from Brazil," said Guillermo Velázquez to me. We were at a midmorning pause during one of the days the Paraguayan Executive Directorate visited the dam for meetings. That day, an engineering student from the United States was touring Itaipú, and I had been asked to join her as a host. But, in the midst of excited conversations about the hydroelectric potential of the Paraná, Velázquez returned to the economic potential of Itaipú. "I think we will be able to sell Paraguayan energy in Brazil," he said. "What I hear from the business sectors, they are very interested in this. People call me and email me and tell me they are open."

Intimate meals and other forms of socializing offered opportunities to pitch the proposal that Paraguay sell Itaipú energy in Brazil via ANDE and that the rate for ceded energy be increased. Executive Directorate representatives traveled to Argentina and again to the neutral ground of Spain to

FIGURE 3.2. Balmelli (right) speaks to industry leaders in Asunción. In the front row, from left to right: the Paraguayan Itaipú director of finance, Vice President Federico Franco, President Fernando Lugo, unidentified businessman, and Ambassador Eduado dos Santos. Source: Author's photograph, July 2009.

make their case. Even in the final weeks of negotiations—a state visit by Brazilian president Lula da Silva had been planned for late July—the Itaipú technocratic team continued to host Brazilian diplomats and energy sector leaders, not counting on the burgeoning presidential momentum to see the agreement through. Just weeks before the signing of the Joint Declaration, the Paraguayan Executive Directorate presented to industry leaders from Paraguay and Brazil at the inauguration of the Itaipú booth the Paraguayan Expo, an event important enough to merit the attendance of President Fernando Lugo, Vice President Federico Franco, and Ambassador Eduardo dos Santos (see figure 3.2).

For months, as the Hydroelectric Commission met in school buildings and pressed for renegotiating the treaty, selling energy to Chile and Uruguay, and renouncing the construction debt, the Itaipú Executive Directorate team visited business and political leaders in South America and Europe, presenting the argument to enforce the treaty and allow Paraguay's ANDE to commercialize its energy in Brazil. But the argument did not end there. In multiple trips to Brazil and in many Asunción meetings with Brazilian representatives, a particular vision of what this closer economic integration between Paraguay and Brazil would look like emerged: ANDE would sell energy in such a way that the end consumer would notice no increase in price. A second part of this newly proposed business partnership included the creation of a *maquila*-type industrial center in Itaipú, on the Paraguayan side of the river, where Brazilian industries would set up factories to take advantage of cheap energy and labor costs.

Unlike the Hydroelectric Commission, the team of technocrats in Itaipú's Executive Directorate acted as if change in Itaipú were best achieved through the day-to-day of administering the dam and through agreements fought out in the cordial but competitive relationship between the two halves of the board as they oversaw the dam. While they were more than willing to bring the debate over Itaipú to an international audience any chance they could especially via the international press, it was always with an expressed concern to not come across as hostile.[22] And after word of a particularly difficult interaction at a binational board meeting hit the presses, Balmelli was quick to publicly state, "Samek is my friend." [23] I suspect that, if they could have, the team of technocrats in Itaipú's Executive Directorate would have entirely circumvented a public debate about hydroelectric sovereignty by merely building the consumption capacity of energy within Paraguay to such a degree that it used its half of the electricity. In fact, though they readily used the Six Points Memo terminology of "unrestricted access" and "fair price," they refrained in public and in private from using the phrases "hydroelectric sovereignty" and "renouncing spurious debt." And so, rather than framing the relationship with Brazil as an imperialist one where the goal would be to leave previously made obligations, the Itaipú technocratic team characterized the relationship as a partnership, where the goal would be that it continue and deepen. With no Paraguayan ambassador to Brazil and with halted formal conversations between the two foreign ministry teams, the Itaipú Executive Directorate assumed the function of Paraguayan foreign diplomacy.

Conclusion: Signing the Joint Declaration, Sowing the Seeds of Discord

All seemed to be going well until the end of July. Even though the Itaipú proposal had been stewarded by dam personnel, diplomatic protocol took over as a list of agreements was finalized. The Brazilian Foreign Ministry formally presented the Itaipú agreement to the Paraguayan Ministry of Foreign Relations in anticipation of the presidential visit by Lula. Unlike the first Brazilian offer of January 2009, this proposal had been worked out in close consultation with Paraguayan counterparts, making acceptance much more likely. Paraguayan foreign minister Hector Lacognata, Ricardo Canese, and ministry members met with their Brazilian Foreign Ministry guests for days to hammer out the last details.

In the midst of the percussive fanfare of camera flashes, Lugo and Lula signed the Joint Declaration on July 25, 2009. The thirty-one-item document resolved the diplomatic disagreement between the two countries, fulfilling many of Lugo's campaign objectives. Joyous crowds flooded the streets of Asunción. With an expression of relief, Domingo Laíno declared that the agreement signaled the end of the War of the Triple Alliance.[24] Ricardo Canese was more guarded, saying, "At this stage, this is very important. . . . But, based on what it addresses, this is not the final stage of negotiation, rather, an inflection point because it only addressed the least important Points."[25] Yet when Lula's entourage arrived and met with the Paraguayan leadership, when Paraguayan crowds cheered, when photo ops at the presidential palace showed Brazilian and Paraguayan government officials shaking hands, one group was markedly absent: Carlos Mateo Balmelli or any other member of the Itaipú technocratic team. After months of behind-the-scenes work by the Itaipú Executive Directorate, the Hydroelectric Commission publicly celebrated the agreement.

Fernando Lugo's coalition government, necessary for electoral victory, had poorly stitched together unity (and tightly knitted antagonism), which became evident in the Itaipú negotiations. The Itaipú technocratic team successfully drafted a resolution to the diplomatic conflict initiated by the leftist opposition, indicating which logics seemed to work on an international platform and suggesting that some techniques that got the Left into national office at times worked against them in international diplomacy. But without the decades of activism from the left, without Lugo's and Canese's insistence to center hydroelectric sovereignty, without the ongoing public pressure of ordinary Paraguayans who were convinced by the progressive narrative, it is doubtful any renegotiation would have happened at all. Indeed, the two governments succeeding Fernando Lugo (a brief Liberal and a full-term Colorado administration) returned to the nondisruptive rhetoric of the past. Lula's leftist sympathies were key to making closure possible.

Drastically different ideas of the nature of state power and differing networks of partisan patron-client loyalties became evident in the way that the two energy teams went about negotiating with Brazil. Because of distrust within Lugo's governing coalition, including between the two Paraguayan energy sector teams, a major fissure in the alliance erupted, a breakdown that led to the 2012 impeachment of Lugo and the advancement of his Liberal Party vice president Federico Franco. The Hydroelectric Commission and progressive sectors in Paraguay described the state as something that needed to be captured and occupied (and thus purged of the contamination from those who had been in it before) in order to turn it away from what it had been—a tool of the oligarchy and bourgeoisie used for extracting wealth from the people. The solution was a Bolivian or Venezuelan participatory democracy to force the state into what it should be, the people. The foe described in Brazil was the foreign service, which had a school that trained the diplomatic corps to be brutally effective negotiators for Brazil's imperialist aims, an example of a state yet to be fully transformed by the people and thus still serving the extractive goals of the elite. Once this was accomplished, a Great Homeland would unite all the peoples of Latin America, not divided into the antagonistic and self-interested nation-states of the present.

For decades, according to this narrative, the Paraguayan state had given in to the imperial interests of Brazil, which it reinforced through pay-offs, through a well-groomed diplomatic corps, and through threats of military intervention, a continuation of how the occupying forces of Argentina and Brazil restructured Paraguay in their interests after the War of the Triple Alliance. The lived experiences under the Stroessner government of Tekojoja's leadership, Canese, and the other members of the Hydroelectric Commission were consonant with this. Thus the state they saw was one of brute violence and physical conflict. And the international state system existed to facilitate imperialistic predation, mostly by Brazil, though the United Kingdom and the United States lurked not too far behind. These sectors framed their arguments, intended to sway the masses in Brazil to sympathy toward the Paraguayan case, as moral appeals about dictatorships and injustice suffered by Paraguayan people. The state imagined here was an illegitimate force acting in the interests of international capital that had to be occupied by the people and turned to serve the interests of the worker and the campesino. If the problem in Brazil was a predatory imperialist apparatus, then the solution was mobilization by the Paraguayan and Brazilian people. And since the corollary problem in Paraguay was that the people were not educated enough about the situation, the solution was to raise their awareness and mobilize them.

On the other hand, the ways that the energy, law, and finance experts in Itaipú's Executive Directorate talked about and then implemented a strategy for dealing with Brazil revealed another conceptualization of how power was distributed within the nation-state apparatus and within the international state system. The technocratic team engaged with a state that was primarily a system that operated under certain rules, and so the question was how to use the rules of the game to gain an advantage. Balmelli's background as former leader of the senate and candidate for vice president, as lifelong active member of one of Paraguay's two chief political parties, as an attorney, and as a doctor of political science shaped the kind of argument he thought would be convincing and permeated the language he used: enforce the treaty, exercise sovereignty, integration and leadership versus unilateralism and hegemony. The rhetorical turn to respect the treaty implied a confidence in the efficacy of law to lead the direction of the state. His closest advisers were engineers with credentialed knowledge of international finance and energy markets who drew up earnings projections and electrical infrastructural blueprints. Thus, the state was a matter of leadership and law. The central problem to be solved in Brazil was that its leaders in and outside the government had not been convinced by the Paraguayan clamoring for unrestricted availability because they had not yet presented a rational solution coming from a serious government that was based on economic and legal precepts.

Paraguay's problem, then, was that its government was disorganized and its arguments not rational. Whereas the Hydroelectric Commission positively mentioned Bolivia and Venezuela, the Itaipú team regularly invoked Lula's leadership in Brazil: one that fostered economic development and social responsibility without drastically threatening the financial model established during previous governments. By presenting legal arguments from the Latin and spreadsheet earnings projections, inviting an integration based on the establishment of industry (primarily Brazilian) within Paraguay, and taking advantage of its competitive advantage—always with the counterexample of the United States' greedy foreign policy—the technocrats within Itaipú attempted to woo key Brazilian decision makers. Rational (as opposed to moral) arguments and the allure of economic expediency were directed toward business elites in São Paulo and politicians in Curitiba and Brasilia, a targeted intervention at the heights of power.

Because the arguments advanced in popular settings in Paraguay, however, did not reach the same level of saturation in Brazil, a double discourse regarding Itaipú arose: one, effective internally to Paraguay, centered on recovering hydroelectric sovereignty lost because of rapacious Brazilian imperialism and weak-willed Paraguayan turncoats, with popular mobilization imagined as the key trigger for Brazilian acquiescence; the other, an external discourse mobilized an entirely different repertoire of arguments and knowledges. The very fact that the Itaipú technocrats used the terminology of the Six Points Memo marks the success the Hydroelectric Commission had in shaping the national energy imaginary, a sign of their expertise in the area, gained from decades of social mobilization. The line of argument and public performance of Balmelli and his team invoked law, physics, and economics—based on an expertise gained by having worked in the public sector or industry for decades. As a pragmatic approach to Brazil, by avoiding accusatory language and threats of renouncing the debt or the treaty, or of involving external arbitration, it deescalated the tension. The public performances were not merely show; they were strategic choices from two coalitions within Paraguay whose attempts to secure Paraguay's interests in Itaipú were indelibly tied to ideas these two groups have of how states should and do work. The question of performance in state-making gauges intended audiences, permitted and prohibited scripts, and stage setting, and from these, it draws conclusions about strategies in achieving the goal of governance.

The raised stakes in the negotiation—an opportunity to recapitulate Paraguay's history, to "win" the War of the Triple Alliance—meant that the jockeying over the negotiations within Lugo's government took on greater urgency. Itaipú's impact, however, came from more than its role as a significant source of income for the central government and of public employment. Itaipú was important because the changes in and around it were also about Paraguay's relationship with its most important neighbor, a bilateral relationship situated within regional relationships. Transforming the way Paraguay acted toward the international community and the state's options for action in this arena fed back on how international organizations (the UN, Mercosur, World Bank) treated Paraguay. The bilateral water issues were seen as precedent setting throughout the continent. This was, perhaps, Paraguay's greatest point of leverage.

To return to questions about the Latin American leftward turn posed by the scholars at the beginning of this chapter, the distinctions between the Hydroelectric Commission and the Itaipú Directorate are not just those between an old versus new left or a right versus left. Both groups advocated state intervention, rejected privatization, and featured critiques of imperialism to varying intensities, with calls for integration at different scales (popular versus economic). And they embraced extractivist market mechanisms as a liberatory path forward. The electoral-campaign-turned-international-campaign was not so much a rediscovery of indigeneity or the reaffirmation of marginalized cultural forms. Instead of identity politics, the momentum behind Lugo's campaign arose out of vast economic discontent. Were it not for the bitter infighting that tore at the Lugo coalition government, it might seem that the progressive administration happened on a brilliant and successful strategy. Two simultaneous negotiating teams working in tandem, playing to their own strengths, kept public pressure alive in Paraguay and found the right inducements and interlocutors within Brazil. The approaches from the two corners of the energy sector arose because of more than just fortuitous chance. Rather, transnational resources lend themselves to creative (and sometimes destructive) political experiments. The progressive Hydroelectric Commission, with a view of the state as illegitimate, needing to be redirected by sovereignty-expressing occupation from below, kept the issue at the top of government. But it was a legalist view of the state, where sovereignty was demonstrated through an international presence and where international pressure from above was the key pressure point, that scripted the resolution.

4

Debt

Introduction

Even though no one else was in the meeting room, Henry Winters lowered his voice to a conspiratorial tone. "The real business of Itaipú," he said, "is debt."

After hours of carefully explaining the mechanics of hydroelectricity in Itaipú, the Anglo-Argentine engineer charged with orienting me on the dam finally got to the twists and turns and underpayments by which the $2 billion construction debt had grown to $60 billion. State secrets, natural resources, and the lurid side of integration were combined in a story of how Brazil's electricity sector influenced the Itaipú leadership team to set the tariff lower than what was needed to pay Itaipú creditors, chiefly Eletrobras. He added, "It suited Itaipú to operate at a loss because the more the debt increased, the more money they would be making." He never said whom he meant by "they," an uncharacteristic vaguery on the part of the precise engineer.

His words were echoed a year later in a conversation with Patricio Delgado, a successful Paraguayan businessman who split time between Asunción and Paris. Knowing that my research project examined governance in Itaipú, he apprised me of his analysis. "When I first wrote [my thesis] about Itaipú, I thought it would bring development and wealth to Paraguay," began Delgado. "But now, I think that they are never going to pay off the debt. 2023 will come, and Eletrobras will say 'pay what you owe,' and Paraguay will not be able to afford it and will have to sign over ownership of the dam to Brazil."

Given these alarming reports about 2023, the expiration of the finance-related terms in the treaty and the date all loan agreements would be due—from cosmopolitan, business-savvy voices and given the prominence of the debt in the Hydroelectric Commission renegotiation goals—I repeatedly inquired about the debt during interviews and weekly visits to Itaipú headquarters in Asunción.

Unsurprisingly, the Itaipú executive presented a different view of the debt than the Hydroelectric Commission Foreign Ministry negotiating team. I met with Esteban Solanzo in his large office on the fourth floor, the Executive Directorate. Because Solanzo possessed decades of senior-level experience in the Paraguayan electricity sector, I asked him about the Itaipú debt.

"Everyone I ask here says it's being paid, but then in the newspapers and in presentations, I hear that it is not being paid," I said to Solanzo in March 2009.

"It is being paid. That is a lie," he replied emphatically.

This was the only time during our conversation that Solanzo raised his voice; the rest of the time he was jovial, happily describing his new iPhone as "amigable," user-friendly. Solanzo was a senior member of Paraguay's electricity sector, helming ANDE in the past and now advising the Paraguayan Itaipú Directorate in its negotiations with Brazil. He, like Guillermo Velázquez, was an engineer with decades of electricity sector experience and formed part of the technocratic team gathered by Itaipú executive director Carlos Mateo Balmelli. It seemed that Solanzo took the question almost personally, a response likely explained by the fact that he had served on an Itaipú debt negotiating team in the 1990s. He reached for a sheet of paper and, with determined strokes of a pen, sketched a diagram explaining the Brazilian energy sector (a map that even included high-voltage transmission lines) and a series of line graphs accompanied by equations to demonstrate the payment plans before and after the inflection point of 1996. Solanzo was exact, noting the curvature of lines, breaking down the tariff formula by its components, and using the technical term "unit cost of electricity service" (*costo unitario del servicio de electricidad*).

Debts of all kinds construct sociality, even between states. And in the case of the Itaipú debt, the bank notes and the stories of obligation laid the foundation for binational integration for years to come. The debt dominates how Paraguayans complain about Itaipú and their anxieties, to the exclusion of other challenges, such as the limited 150-year operative life span of the dam and ever-increasing energy consumption in Paraguay with no plan in place to source electricity after capacity is reached (which ANDE places as early as 2023).[1] A topic of frequent controversy among hydroelectric dam projects worldwide, debt financing often links national economies to international creditors in troubling ways. In some cases, the life of the dam is even exceeded by the life of the debt (as appears to be unfolding in Paraguay's other binational dam, Yacyretá). As of my 2009 conversation with Solanzo, Itaipú had sold $46.5 billion worth of energy, and, since about 64 percent of the revenue went to debt payments, the dam had paid its creditors $30 billion.[2] Itaipú alone was responsible for the construction debt, which does not appear on the national ledgers (making the Brazilian and Paraguayan governments appear less indebted). Over the first twenty-eight years of energy sales, 95 percent of that $46.5 billion came from the Brazilian market. According to the Brazilian executive director (reporting on Itaipú financials at an extraordinary session of Mercosur), from 2009 to 2023, the dam was expected to pay another $28 billion to its creditors before fully discharging the debt.[3] Since the majority of the energy was sold to Brazilian consumers, the majority of the revenue, including what would be "Paraguay's half," came from Brazilian consumers. Yet, while the compensation paid to Paraguay was disputed, there was no area more contentious than the debt—so much so that, during the Lugo government renegotiations, the Brazilian government was willing to negotiate and even yield on the price paid for energy, but it was not even willing to discuss reevaluating the debt.

Like the tariff, the debt of Itaipú was bound to national histories and even the national character of Brazilians and Paraguayans. The origin stories of the debt recounted by energy technocrats, leftist activists, and concerned citizens had grown into a bramble-like confusion for reasons beyond just the expected mystification around complex numbers or the persistent national resonances within a binational institution. All these origin stories seem to grapple with obligation, on who and what constituted morally obligatory behavior. If the resources from the tariff were to bolster national development, safeguard the sustainability of energy generation, and recognize the sovereignty of both parties, the debt continued to be the site of a fierce standoff over national sovereignty. All this was exacerbated by the juridical/accounting isolation of the dam, which occulted financial decision making. Because the numbers on the Itaipú books were not examined by any outside agency or made public (until 2010), the debt lent itself to conspiratorial fantasies. What the debt was for, that is, its purpose, depended on the position of the speaker relative to the dam. To a seasoned administrator like Solanzo, it was the necessary cost of doing business; to Delgado it was a five-decade strategy to take full control of the hydroelectric potential of the Paraná River. Because of Itaipú's scale-producing qualities—its dominance in the energy market and its binationality—debt converts energy into state power, suggesting that more was in play than any single constituency's representation of the dam's debt.

Mirroring the energy infrastructure, the financial infrastructure was unevenly distributed within Itaipú. Compensation, a rent-like and not market-based amount, linked the dam and the Paraguayan state. Compound interest, on the other hand, characterized the financial architecture between Itaipú and the Brazilian state. Both were successful accumulation strategies. Interestingly, rent-based growth was criticized as fiscal negligence, but the spectacular debt-based financial arrangement got presented as a virtue of risk, only recently drawing outside attention. Notwithstanding language of a cooperation that superseded national identity, narratives about the morality of the debt reinforced nation-based representations. As others have attended to the almost theological valence of beliefs about debt, competing moralizations of the Itaipú debt elided into the moral qualities of Brazilianness or Paraguayanness.[4] But the transformation I am most interested in is the sublimation of debt moralities into the nigh spiritual qualities of the nation-state. By tethering virtue to national community to national state, representations of the Itaipú debt reinforced the symbolic power of the Brazilian and Paraguayan national states.

The Itaipú debt is in the context of other local experiences of indebtedness. A thriving debate in Latin Americanist social science on money, finance, and value offers much to flesh out how debt and circulations work in day-to-day life, an important contrast particularly in light of how frequently Latin American state debt is a topic of study. Although he intended to study political clientelism among marginalized communities in Argentina, Ariel Wilkis found that tracing moral hierarchies in money instead allowed him to see how the various social worlds of the poor connected and operated. From his ethnographic research, he frames money as "pieces of money" that are "shaped by ideas and beliefs about morality," and so "each of these pieces of money differs from the others."[5] Thus he notes differences between "lent money" and "earned money," "political money" and "sacrificed money," using money as a conceptual tool to trace social connection. Since, as Wilkis asserts, all money is moral and is internally composed by value hierarchies, we should expect to see that different Itaipú energy monies have different moral valences: energy rent, hydrodollars, energy debt. For example, in much of Paraguay, the social funds spent by Itaipú are positive whereas the debt (as we will see in this chapter) has a vicious moral economy. Moreover, because "no single piece of money possesses a single meaning,"[6] it should come as no surprise that these same monies have different meanings when handled by different actors (e.g., Solanzo and other energy sector technocrats versus the Hydroelectric Commission).

Paraguay, in particular, lends itself to intricate studies of financial flows and obligation, from the colonial yerba mate harvests that generated conflicts between Spanish colonists and Jesuit missions to the post–War of the Triple Alliance bond offerings (arranged by British banks, who also held much of the war debt on all sides of the conflict) to the complicated personal and family finances of the present under pressures triggered by an economy structured around export-oriented soy and beef. Caroline Schuster's work on gender and microcredit in Ciudad del Este, the Paraguayan border city known for arbitrage, triangulation, and a vast informal economy oriented toward Argentine and Brazilian consumers, shows how obligation and sociality are constructed via the circulation of credit.[7] She describes how "social collateral" works within the "social unit of debt" of gendered and gendering microcredit where groups of women, rather than individuals, rely on their networks in order to acquire loans that go to pay for school fees, holiday presents, and the goods necessary for the various small businesses they run out of their homes. The ability to take out a loan is predicated, not on assets or income, but on relationship. Risk and liability are thus embedded in ordinary relationships for many Paraguayans, producing tension and intimacy by alternating turns. But whereas in personal life debt is a form of intimacy with neighbors and extended kin networks, the joint debt in Itaipú enmeshes Paraguay with its former foe in a fraught asymmetry that invokes the War of the Triple Alliance for many Paraguayans. The Itaipú debt shows what happens to obligation and intimacy when they are inescapable and riven by tragedy.

Energy debt needs to be carefully distinguished from petrodollars, hydrodollars, and energy rent. Sandy Smith-Nonini has recently argued that energy debt and the loss of energy sovereignty are interlinked in creating the neoliberal era.[8] By tracing the shift from Keynesianism to neoliberalism in the United States through petroleum production, she therein finds changing formulations of hegemony. Prior to the 1970s, the United States geopolitically dominated Middle East oil reserves, assuring industry a ready source of cheap energy. But the emergence of OPEC, the peaking of conventional domestic US oil production, and the oil price shocks compromised the previous arrangement and facilitated the new economic hegemony of the Washington Consensus as a way for corporate interests to maintain power. In a subtle argument that works against the grain, Smith-Nonini asserts that, because neoliberalism comes out of a need to recover hegemony after losing oil control, the very tools of neoliberalism are energy related, including the prioritization of credit-finance capital over industrial-corporate capital.[9] Smith-Nonini therefore insists on a wide definition of energy debt: the increased reliance on debt financialization in the developing world via the IMF and World Bank, the increased use of debt by corporations, and even the growth of personal household debt.

David Graeber, in *Debt: The First Five Thousand Years*, uses ethnography to help overturn a common economics origin story about money.[10] The conventional narrative of barter in primitive communities leading to a common universal form of exchange (i.e., money) and then, finally, credit and other abstract forms simply has not been documented by a single ethnographer. It also does not cohere with the archeological record. From this, Graeber argues that the historical order of events started with credit-based exchange (first in the form of neighborly reciprocity and then overseen by imperial governments), and only later did coinage develop specifically to fund the lifestyles of large standing armies. That is, money comes after debt.

Taking these claims as a critical point of departure offers a provocative argument regarding the Itaipú hydroelectric dam and renewable energy more broadly. In chapter 1 we noted the circulation of current and how that accomplished both material duties (in powering industry) and sociopolitical ones (identifying/defending national sovereignty). In chapter 2 we looked at how current becomes currency, how money is made by the dam—most of which is determined by administrative declarations and not a market. Chapter 4 uncovers the development of the Itaipú debt. The order of chapters makes it seem that the megawatt hours of current are primary, leading to multiple tariffs by which the product makes money for its owners, and, thirdly the debt is a consequence of fiscal mismanagement and budgetary needs. But if the primacy of credit holds true—not just chronologically, as it does in Itaipú (loans preceded energy production), but ontologically—then we might understand Itaipú according to Winters's claim that the real business is debt. In fact, to take the matter one step further: is energy money?

Electricity is, after all, the circulation of negatives. Perhaps given the importance of energy to industry and the transnational management of hydraulic resources, different kinds of money circulate on a topographically different scale than those currencies that circulate on a nation-state scale (see chapter 6). Like carbon credits and as-yet-inchoate ecomonies (designed to value environmental preservation), the exchange here has to do with a negotiated dissatisfaction with untempered fossil fuel consumption. This suggests that money may take different forms, not just appellations ("pound," "yen"), depending on geographic scale. In a situation of primary dependence on renewable energy resources and amid new political-economic adjustments forced by the Anthropocene, energy itself plays multiple roles, a hint of which Itaipú forecasts.

Public debt goes to the heart of political and economic power.[11] Perhaps this is why it was so difficult to research. Folk theories about Itaipú finances were easy to gather; they persistently appeared in op-ed pieces, in interviews, in dinner conversations, no matter how many times they were refuted by Itaipú technocrats whom I shadowed. But the actual numbers, the chronology of debt agreements, the promissory notes were much harder to find because they were either legally or pragmatically unavailable to the Paraguayan or Brazilian publics. Prior to the Lugo government in Paraguay, all Itaipú financial records were kept within the dam, outside of the jurisdiction of Paraguayan state institutions. Only minimal, PR-sanitized figures on income and energy production were published widely. And of the two energy negotiating teams, only one group (the Hydroelectric Commission) emphasized the debt, in both public and private.

Chapter 2 broke down the tariff formula into its component parts. In this chapter, I begin with the historical development of the debt and the transformation wrought on state institutions as Eletrobras metamorphosed into a creditor, uncovering how the multiplex nature of state institutions is connected to renewable energy requirements. A "multiplex" is a channel or circuit along which multiple signals are simultaneously sent, an analogy from telegraphy that applies to the Itaipú debt, which I argue is a multiplex along which multiple and often contradictory messages are transmitted at the same time. After piecing together a biography of the debt, I will turn to a more ethnographic exploration of how Itaipú debt origin stories function based on ideas of "interest." Christopher Gregory urges his readers to heed Strathern's work of economic anthropology *No Money on Our Skins*, which explored economic transactions between kin and close friends in Papua New Guinea.[12] To understand money lending between family and friends, Gregory reminds us, required more than a peak at the books; it necessitated gathering family histories of closeness, previous assistance, and betrayal. It turns out that this is also the case for the debt of the largest dam in the world.

The commitment to place from chapter 1 that so strongly influenced the mechanical infrastructure of the dam has ramifications in the finances of Itaipú. The infrastructure of renewable energy serves as collateral for the debt, state-owned collateral, in fact. So does the hydroelectric potential. And neither can be easily repossessed by displeased lenders. As a result of the engineering decisions of chapter 1 (namely the choice of local consortia for the construction), much of the debt circulated locally. What this means about the political ecology of Itaipú renewable energy is that the debt, because so much of it was state held, strengthened the institutional and financial infrastructure of the Brazilian nation-state (and the Paraguayan nation-state, to a lesser extent). Some parts of the state apparatus grew in size and density; others, like Eletrobras, underwent a fundamental qualitative change in nature. And so, because the debt simultaneously transmits several narratives and the financial infrastructure simultaneously hosts the accumulation strategies of rent and compound interest, renewable energy finances foster political-economic multiplexity.

A Biography of the Itaipú Debt

Construction began shortly after the Itaipú Treaty was signed in 1973 and continued for almost two decades. At its peak, more than thirty thousand people were directly employed by the effort, including Henry Winters. Mirroring the construction, the debt that paid for it may be divided into three phases. The first was a lengthy construction period without energy production (i.e., no sales to be able to begin paying off the debt), planned to run from 1973 to 1983, but in actuality extending from 1973 to 1984–85. The confusion on that latter date is critical and one that will be discussed below. The second period was construction with energy production as individual turbines were brought online (i.e., sales that would bring income to begin paying off debt), planned for 1983–89, in practice 1984–85 to 1992. The final phase would occur once the dam was complete and fully operational, with debt amortized to 2023, the fiftieth anniversary of the Itaipú Treaty, if not beforehand. Over that time, Eletrobras moved from being guarantor of Itaipú's loans and anticipated vendor of the electricity to being a lender, as well.

TABLE 4.1. Total Loaned to Itaipú, 1974–2008

Year	Amount, US$ Millions	Year	Amount, US$ Millions
1974	5	1992	1
1975	180	1993	1,893
1976	242	1994	82
1977	598	1995	78
1978	1,004	1996	137
1979	1,234	1997	483
1980	1,457	1998	48
1981	1,891	1999	0
1982	2,012	2000	53
1983	1,360	2001	14
1984	1,739	2002	74
1985	1,685	2003	68
1986	1,658	2004	30
1987	2,048	2005	23
1988	1,733	2006	12
1989	1,854	2007	21
1990	1,840	2008	8
1991	1,386	**TOTAL**	**26,951**

Source: Parlasur 2008, PowerPoint presentation by Jorge Samek.

When Itaipú was still being planned in 1972 and 1973, one of the first financial decisions taken was that, although the dam was equally co-owned by Brazil and Paraguay, Brazil alone would be responsible for securing all the funding for the dam. At that time, the total estimate for construction costs, including servicing of the debt, was a staggering $2 billion. Over the next four decades, the amount borrowed by the dam grew to $27 billion (see table 4.1). In 1973, the GDP of Paraguay was less than $1 billion; the GDP of Brazil was $79 billion in current US dollars.[13] The $2 billion price tag was shocking, hitherto unheard of in the megaproject world, and just another sign of how gargantuan the binational dam would be. Eletrobras, understood to be a proxy for the Brazilian government, provided all the guarantees and collateral and negotiated with creditors. Paraguayan officials in Itaipú merely signed the loan agreements. (The debt was held entirely by Itaipú, even if the funding was secured by another entity.) And so, from the very beginning, risk was not distributed evenly within Itaipú.

Various explanations have been given for this arrangement. The first Paraguayan executive director, Enzo Debernardi, recounted that, "by a decision rooted in the government of Paraguay, all its objectives were bound to the principal nonnegotiable condition to not support the undertaking, either through financing or guarantees, in order to reserve all possible lines of international credit for other projects related to the country's development."[14] With a much smaller economy than Brazil, argued Debernardi's team in 1973, contributing collateral or foreign exchange to secure any part of the massive debt would circumscribe Paraguay's ability to embark on other debt-funded development projects. But when the Brazilian delegation was informed of the decision, their response was of "surprise and nigh incredulity," and they attempted to negotiate a common ground where Paraguay might support even just a fraction of the financing.[15] The Paraguayan team remained resolute. Though Debernardi's memoirs mention the nonnegotiable posture of the "government of Paraguay," this phrase at the time was a euphemism for but one thing: the person of Alfredo Stroessner, unquestionable ruler of the country. If the government of Paraguay took a position, there was no hope to shift it. Debernardi and other Paraguayan negotiators communicated the stance to their Brazilian counterparts, who acceded in order to move forward.

Henry Winters was less circumspect when speaking to me. "Stroessner did not want the dam to cost Paraguay one cent," he said as he explained the contours of the initial debt planning. And, in early 2009, in the thick of international disagreement over Paraguayan requests to get more money for exported electricity from Brazilian consumers, Brazilian energy minister Edison Lobão used the same language to express in public what many were saying off the record: Paraguay did not contribute "even one cent" (*um centavo sequer*) to construct Itaipú.[16] Ostensibly, the Paraguayan position was for the sake of development, but the critique embedded in the statements of Winters and Lobão sees self-denial as a virtue and thus ascribes the moral high ground to debt. Debt may be a sign of community membership; if so, its absence is a sign of exclusion.[17] The social advantages of the Itaipú debt quickly accrued to the government of Paraguay, bringing Paraguay into relationship with the international community. Paraguayan newspapers carried advertisements announcing further lines of credit extended by foreign banks and foreign firms after the Itaipú project commenced. The new lines of credit allowed the Paraguayan government (read: Stroessner) to develop other pet projects, usually construction oriented, like the paving of major roads, much of which simply translated into direct cash transfers to Colorado Party elite. The new liquidity transformed daily life in Paraguay even as it created a new middle class and service sector. Thus, without actually holding any of the debt of Itaipú and without contributing collateral,

Paraguay still benefited from the international legibility brought on by the debt.

Eletrobras officials on the Itaipú Governing Council and the Itaipú Directorate turned to a number of sources to secure all the financing: buyer's credits with equipment suppliers, banks in the United States and Europe (notably Citibank and Deutsche Bank),[18] the Brazilian state development bank, the Inter-American Development Bank. (The Itaipú Governing Council is an administrative board responsible for long-term planning, a more traditional "board of directors" to the Executive Directorate.) Then the global economic crisis of the 1970s struck, contracting credit worldwide and driving up interest rates. The oil embargo drove up the cost of materials. As financing dried up, Eletrobras began to loan money directly to Itaipú itself. The first $2.5 billion loan from Eletrobras to Itaipú was signed in 1975. As the credit crisis stretched throughout the decade, more and more debt came from Eletrobras. And what occurred was a strange collapsing of identity: Eletrobras, first a kind of debtor as co-owner of Itaipú and guarantor of its loans, became a creditor. By the end of the decade, the total debt of Itaipú exceeded $4 billion.

Electricity sales and income were delayed by two years because energy production commenced later than planned. Turbine construction, particularly, took longer. Buyer's credits and foreign loans moved out of their grace periods in 1983, the original start date for energy sales. With credit constrained worldwide because of anti-inflationary US monetary policy, there was nothing to be done but refinance the debt that was due. But construction costs continued to rise with oil prices and inflation, and so Itaipú relied more and more on money from Eletrobras. Although the base interest rate for the Itaipú debt sat at 10 percent, because of various adjustments, the actual number came out to between 13 and 14 percent, depending on the loan. The debt situation seemed ripe for disaster during the first phase of construction; it was nothing compared to what occurred once the turbines began churning out energy. Two turbines finally generated energy in 1984, and one of two things happened: either Itaipú drastically undercharged for that electricity, or, in fact, it charged nothing. And when sales definitely commenced in 1985, the base cost of energy was woefully insufficient.

According to official Itaipú-published documents, when energy sales began in 1985, Itaipú's Executive Directorate decided to charge US$10 per kilowatt month for the electricity because Eletrobras, the primary market for the energy, argued that the Brazilian consumer could not afford a higher rate.[19] Because of the hierarchical ranking of "executive" directors within the Itaipú Directorate, financial decisions were initiated by the Brazilian Financial Direction of Itaipú. The guaranteed energy tariff rate of US$10 per kilowatt month fell below the minimum necessary to cover operation costs and the interest rates on construction loans—in violation of the tariff formula mandated by annex C of the Itaipú Treaty. Eletrobras's rationale for this low rate was that Brazilian industry was in an exceptionally vulnerable moment (this was the "lost decade" in Latin America, after all) and that anything higher would put Itaipú electricity above the market price of energy. The hope was that the following year, the rate would be raised and more turbines installed, bringing in more revenue to the dam.

During Fernando Lugo's presidency (2008–12), energy production, sales, and accounting documents were examined by Paraguayan government officials for the first time in the history of the dam. And though Lugo was deposed in June 2012, a special audit report on the Itaipú debt was released by the general comptroller of the republic later on that year.[20] The audit report is crucial because it is the first public report ever on the Itaipú debt from an external body. The findings of the comptroller agree with the official Itaipú assertion that electricity was sold to ANDE and Eletrobras at US$10 per kilowatt month in 1985, below the minimum necessary to service the debt. However, the comptroller audit points out that, while Itaipú energy production began in May 1984, Itaipú began reporting income only in March 1985. That is, for nine months, ANDE and Eletrobras received electricity for free from Itaipú or, perhaps, this electricity was consumed by the dam itself. Because Paraguayan access to the Itaipú books is still restricted, there is no way to verify where this energy went or what was done with the economic proceeds.

After the first year of energy sales in 1985, the leadership of Itaipú again debated the electricity tariff. The debt on the books was $9 billion, already up from $4 billion at the beginning of the decade. Estimations of the total amount that would be paid on the debt between 1985 and 2023, including interest, were about double that. But the Brazilian economy was still in shambles, with rampant inflation and job loss. Plan Cruzado, an extreme price-freeze and wage-freeze attempt to tame triple-digit inflation, began in 1986 and foundered by early 1987.[21] Eletrobras officials again argued that Brazilian consumers could not afford the rate proposed by Itaipú technocrats, a rate of about US$16 per kilowatt month, which would cover all expenses including the debt; the tariff remained depressed. It was US$10 per kilowatt month again in 1986, $11.28 per kilowatt month in 1987, $13.38 in 1988, $13.76 in 1989, $14.98 in 1990, and so on (see figure 4.1).

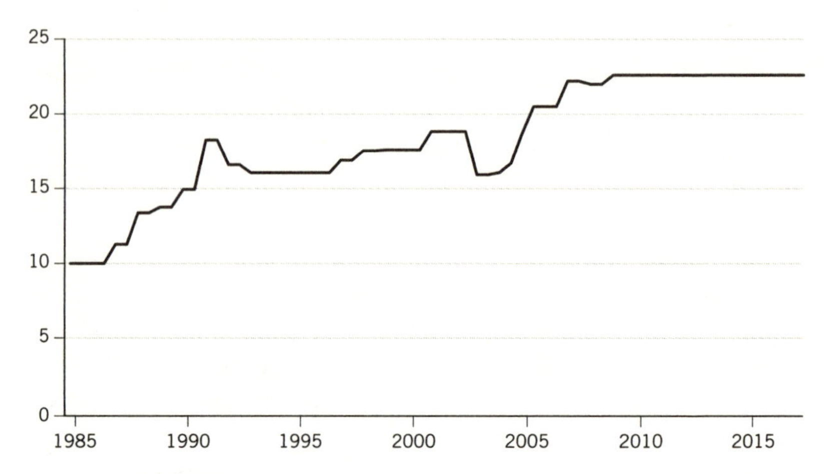

FIGURE 4.1. Itaipú guaranteed energy tariff, 1985–2017 US dollars per kilowatt month. *Source*: Itaipu Binacional, 2003, 2006–2018

And so, even as inflation drove up wages and interest drove up the debt, raising the minimum price per kilowatt month needed to break even, the tariff was intentionally discounted. A loop emerged. Eletrobras sold the vast majority of the energy produced by Itaipú, of which it was a partial owner. It then paid Itaipú for that energy. Itaipú then paid its bills to its creditors, including Eletrobras. But Eletrobras officials were instrumental in setting the price that Itaipú would charge for electricity and insisted that the price be lower than the interest rates for the money it had loaned to Itaipú, ensuring that Itaipú would not be able to fully meet its obligations.

The debt grew, but not equally among Itaipú creditors (see figure 4.2). The creditor whose debt grew the most in the first decade of energy sales was Eletrobras. In fact, the debt to Eletrobras alone grew by almost 400 percent from 1985 to 1995, even though much of the construction was finished. At the beginning of 1985, $4.2 billion (46 percent) of the total principal of $9 billion was owed to Eletrobras. When 1992 started, the first year where all the turbines were in operation, the total principal was $16.9 billion, of which $8 billion (48 percent) was held by Eletrobras. The ascent continued. By 1995, $15.4 billion (89 percent) of the $19 billion debt belonged to Eletrobras. Factoring in interest, the total to be paid to all Itaipú creditors by 2023 would be $60 billion. The slope of the curve continued upward, with no sign of an inflection point.

Faced with the prospect that the debt would never be paid off (let alone zeroed out in 2023, as originally planned), the dam's leadership finally took decisive action and commenced a three-year period of negotiation (1993–

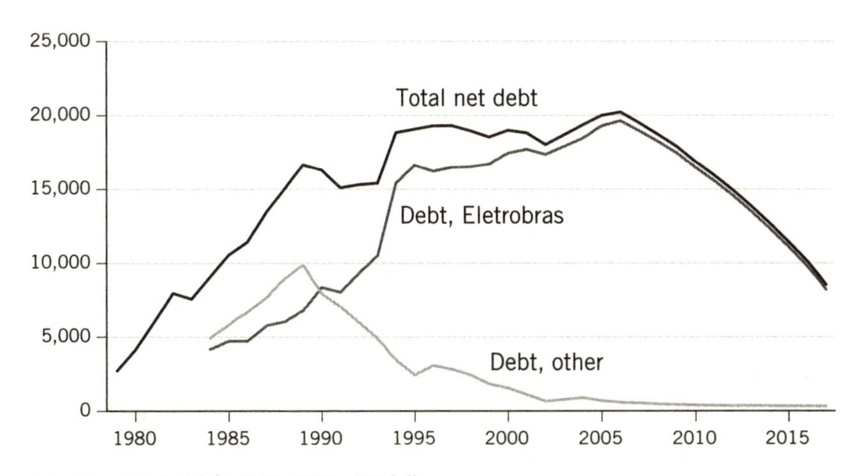

FIGURE 4.2. Itaipú debt, 1974–2017 in US dollars.
Source: Itaipu Binacional 2003; 2006–18.

96). The dam reached a peak crisis in April 1993 when it only received 1 percent of what it billed Eletrobras (this was eventually repaid). Part of the matter was that, although Itaipú operated in US dollars, the Eletrobras debt was in cruzeiros, the volatile Brazilian currency. Payments in relatively unstable local currency with debts held in stable foreign currency are a common problem with megadams worldwide and contribute to their role in "underdeveloping" underdeveloped countries. But in the Itaipú situation, there was a third part to the transaction: Eletrobras paid Itaipú in cruzeiros; Itaipú then converted cruzeiros to US dollars and then, later, back to cruzeiros in order to pay Eletrobras its loans, a form of financial hydrodollar arbitrage that helped Eletrobras. As confidence in the cruzeiro dropped and inflation increased, Itaipú's energy sales were affected. Though Plan Cruzado had failed in the mid-1980s, Fernando Henrique Cardoso's Plan Real was implemented in 1994, introducing a new currency to stabilize the Brazilian economy. It worked. But it also coincided with a jump of $4.9 billion in the debt to Eletrobras in just that year alone.[22]

"[Beginning] in 1993 we negotiated the debt," said Esteban Solanzo as he drew a graph with dollars on the y-axis and time on the x-axis. His personal participation in ANDE's leadership during the meetings lent him facility with numbers and dates, even more than a decade later. "There were three ways to stop the debt from increasing so much that it would be impossible to pay," he said, explaining the logic and the counterarguments. "Raise the tariff—but this punishes [future users]. Ask Eletrobras to pardon the debt—but Eletrobras was no longer just state owned; it also trades shares;

and they said no. Reduce the interest rate to be able to pay it off. And this last is what we did."

In a series of closed-door meetings over many years, the binational Governing Council and Executive Directorate met and hammered out a new deal. There were two direct changes to the debt: first, Eletrobras lowered its interest rate; and, second, all the Eletrobras debt was dollarized to protect the debt from fluctuations caused by changes in the Brazilian currency. The interest rate went from about 14 percent (including adjustments) to 10 percent (including adjustments). In 1996, the tariff was also raised (by 7 percent) to meet the requirements stipulated by the Itaipú Treaty. At the same time, Itaipú agreed to install two further turbines and to finance this with new loans from Eletrobras. For this reason, although debt payments were finally sufficient to pay down the debt, the debt Itaipú owed to Eletrobras continued to increase in the new millennium (see figure 4.2), where the debt held by non-Eletrobras creditors consistently decreases after 1997). Of the $60 billion that Itaipú would pay its creditors, the dam's financial office estimated that $36 billion would be paid to Eletrobras alone between 1996 and 2023.

As Solanzo alluded, the mid-1990s were momentous for the electricity sector in Brazil in other respects. At the height of neoliberal privatizations in South America (the overwhelming majority of Argentina's hydroelectric dams were privatized by President Carlos Menem between 1993 and 1996), the Brazilian and Paraguayan electricity sectors stand as remarkable. Eletrobras, up until that decade entirely government owned, sold a minority stake to private individuals and corporations—a partial privatization—in 1995. There was no talk of privatizing Itaipú as a way to discharge the debt; it was simply too good of a business opportunity, and its binationality acted as a shield. In fact, the only hydroelectric dams in Argentina that were not privatized were its two binational dams: Yacyretá (with Paraguay) and Salto Grande (with Uruguay). It seems that binational institutionality inoculates against privatization, perhaps because international treaties involve congressional approval and not merely executive mandates; and also because Eletrobras shareholders benefitted from the Itaipú debt while being sheltered from the risk of the debt.

The energy pricing in the first decade of production (1985–96) subsidized current consumers at the expense of future consumers. In practice, what this meant was that Brazilian industry was subsidized by receiving electricity for free or for less than the treaty-stipulated rate and that future Brazilian consumers would have to pick up the slack plus compound interest. And even though Paraguay's demand for electricity increased from what it was in the mid-1980s, it still only consumed less than 10 percent of what the dam produced. Brazil is still the market that purchases Itaipú electricity and provides the revenue that pays down the debt, including most of "Paraguay's half."

Another major outcome of the energy pricing was that the debt owed to Eletrobras more than quadrupled, and the amortized total to be paid grew by more than 300 percent. Perhaps a foreign investor or privately owned energy company might have balked at seeing insufficient debt payments year after year, wondering whether there would be a return if the principal plus compound interest more than tripled. Itaipú might seem too risky. But for the Brazilian electricity company, with the backing of the state in the energy sector and with the spectacular growth of the Brazilian economy, which also quadrupled in the same first decade of Itaipú energy production, this was great business. Eletrobras was guaranteed to be paid back, even if the debt had gone from $4 billion to $36 billion, because it enjoyed monopoly control over the electricity market. This is what Henry Winters implied when he said, "The real business of Itaipú is debt." Yes, there were obvious conflicts of interest in that Eletrobras was instrumental in setting the energy price below what would be needed to meet its interest rate, but to whom could anyone appeal? Whatever the technicalities of the law, Eletrobras and the Brazilian government writ large were above it, demonstrating state power and sovereignty. And now that Eletrobras is partially privatized, its stakeholders have a guaranteed returned from a captive market. This is also an instance of hydropolitics—the dependence on territorialized nation-state actors, actions, and institutions on the part of business and industry for their growth.

Interpreting the Debt

In January 2009, as part of their communications strategy, the Hydroelectric Commission hosted a public debate entitled "Itaipú and Yacyretá Are Also Ours" in the large meeting room of Paraguay's national congress. More than three hundred people crowded into the auditorium—women and men, urbanites in suits and campesinos in starched plaid shirts from the hinterland, professionals and students—to attend an hours-long presentation on the state of negotiations with Brazil. Some had traveled the better part of the day at the hottest time of the year. The cross section of the population differed from the political elite who normally occupied that chamber, testament to how the issue brought the margins of Paraguay into the official halls of power. The speakers, including Ricardo Canese, gave updates on the goals of the Hydroelectric Commission and on the progress toward them at a public forum where so many seats were filled that people sat on the floor in aisles and more than a hundred stood on their feet.

Ignacio González, the first presenter, the youngest member of the panel and a member of the Movement toward Socialism Party (PMAS), focused much of his attention on Paraguay's hydroelectric debt. "Auditing the debt," he said, "is a political act." González urged Paraguay to "socialize the debt" owed on Itaipú and Yacyretá Dams and turned specifically to the example of Ecuador's dealings in its external debt as one to follow. Ecuador defaulted on its sovereign debt in 2008, in spite of booming government cash reserves, and then evaluated loan agreements that President Correa termed "odious" and illegitimate, successfully forcing a "haircut" where debts were restructured to $.38 on the dollar.

Domingo Laíno, elder statesman of the Paraguayan left, also spoke at length on the debt, even though the ostensible justification for the evening was the entire state of the negotiation. "The current debt of Itaipú is $19 billion," he said, "even though Paraguay has already paid more than $30 billion." Poverty in Paraguay, he explained, was the fault of "an empire based on finance" (*imperio*, an allusion to Brazil). The former Liberal Party senator continued, "Japan has offered financing at one-tenth the rate of Eletrobras. Therefore, this isn't usury. It's 'megausury.' "

"The Congress of Vienna gives Paraguay the right to annul the [Itaipú] Treaty," Laíno then said, referring to the 1969 Vienna Convention on the Law of Treaties. "Paraguay can do this unilaterally or it can be done bilaterally." Vigorous applause met this pronouncement. The Vienna Convention (1969), of which both Brazil and Paraguay are parties, codified contemporary international law and recognized the rights of states when forming treaties, but the process of annulling treaties was more complicated than Laíno implied.

When Ricardo Canese finally took the stage later that night, several hours had already passed. Although the three previous speakers represented important parts of the Paraguayan left, it was Canese most had come to see. "In Panama there were good negotiators. In Bolivia there were good negotiators," he said. "But it was the people who attained sovereignty." After this rousing opening, Canese described the main goals of the Hydroelectric Commission, including the Itaipú debt. "We want to review the debt," he said, "and only pay the legitimate debt." And then he gave the deadline for the negotiations with the Brazilian diplomatic corps: August 15, 2009. "If there isn't an advance, we have many other options: arbitration, international courts, adopting unilateral actions. Remember, Brazil renounced its external debt in the 1930s."

The Spanish verb *interpretar* is a false cognate for the English. Two of the most common meanings are "to explain the significance" of a thing and "to act out a role" in a work of theater. In Paraguay and Brazil, the Itaipú debt was interpreted in both senses. As various constituencies in Paraguay and Brazil attempted to explain the debt, how it grew, and what was happening to it, nationally constituted and professionally inflected dramas were acted out. Like a script directing the action of a play, the meanings of money determined what it did (and not just vice versa). The divergences in the public accounting of the Itaipú debt enabled these dramas and interpretations. The common representation of the Itaipú debt in Paraguay was as Brazilian imperialism, a spurious violation of the treaty; in Brazil it was the righteous burden of having assumed great risk; among energy sector technocrats, it represented financial exigency; and to suspicious minds in Paraguay and beyond, the debt kindled any number of conspiracy theories. In interpreting the debt, my informants themselves theorized what state debt was for and, consequently, what could and should be expected of the Itaipú debt. They explicitly linked debt, risk, collateral, and responsibility to nation-state and energy in different and consistent ways. The fact that these understandings of state obligations to citizens or to creditors were simultaneously acted on by different groups empowered to make wide-reaching decisions (e.g., the Paraguayan executive, the Brazilian executive, energy sector technocrats) demonstrates how and why renewable energy finances are multiplex.

As part of the demand for hydroelectric sovereignty, Paraguay's leftist negotiators called for inspection and correction of the spurious debt of Itaipú. In the eyes of Lugo's supporters, to rank-and-file members of other parties who did not support him, and to the popular media in Paraguay, Itaipú's leadership had violated the treaty by charging less than the treaty-specified minimum for the base rate of energy. Moreover, it was well known (despite Itaipú silences) that the price cap had been set at the request of the Brazilian electricity sector and that the creditor whose debt had ballooned most was Eletrobras. The debt, then, was Brazil's fault. And for Paraguayan energy to be sold to pay off that debt was a violation of Paraguayan sovereignty. In a manifesto published even before Lugo became president, Canese termed the debt "spurious, illegitimate, and even illegal" and argued that Paraguay should refuse to pay it.[23] Years later, in the midst of negotiation with Brazil, Canese's close team reiterated the point at the Asunción public assembly.

Canese, Laíno, and other members of the Left regularly gave presentations on the state of Itaipú negotiations throughout Paraguay in 2009 where language decrying the "unfairness of the treaty" and "Brazilian imperialism" was common, to the consternation of many in Brazil. Arguably, these public rallies in Paraguay were a successful counter to the closed-door Hydroelectric Commission meetings with Brazilian representatives, which ceased, rumor had it, when the Paraguayan team not only demanded that the Paraguayan Office of the Comptroller audit the Itaipú debt, but also intimated that the Brazilian government forgive the debt entirely. As part of a public campaign, repeated throughout my interviews and observation of the leftist activists, to make "hydroelectric sovereignty a national cause," the leftist Hydroelectric Commission representatives wielded statistics, some more precise than others (e.g., as Laíno had said, the debt in 2009 was indeed $19 billion, and $30 billion was the total paid on the debt to date, but 95 percent of that had come from Brazilian consumers, not Paraguayan consumers, and, at the very least, only half of that—$15 billion—was Paraguay's half). Yet the scale of the numbers and the complexity of the financing led to inventive communications strategies. The Hydroelectric Commission's public relations office designed a colorful newsprint cartoon, which they handed out to listeners at presentations throughout the year. With the ordinary Brazilian and Paraguayan citizens dressed in their national soccer jerseys, the cartoon compared the hydroelectric dam to a neighborhood grocery store, and electricity to manioc, the root staple of the Paraguayan diet. In the drawings, the money and amounts used in the analogy were kilograms and Guaraníes, but the sacks of cash acquired by the portly personification of Eletrobras looked nothing like Paraguayan or Brazilian currency. They were clearly US dollars.

"We think the debt is zero," said Miguel Flores to me with a triumphant smile. Flores served as an adviser to the Hydroelectric Commission and had collaborated for years with international governance organizations, including Transparency International. At the time of our interview in December 2009, the Paraguayan Office of the Comptroller was still in the thick of auditing hundreds of Itaipú financial documents and had not yet published even a preliminary report of their findings. But Flores and the Hydroelectric Commission were confident that what would be found was that, had the treaty-specified minimum for electricity been charged, the debt would have been fully paid off. And then Paraguay might appeal to The Hague or another international court for restitution for the Itaipú debt. What this meant in practice was unclear since Itaipú (and not the governments of Paraguay or Brazil) was liable for the construction debt and since the debt in Itaipú was treated as a non-nationally distinguished whole, in spite of how some spoke of it as having Paraguayan and Brazilian halves.

Flores, Canese, Laíno, and many other Paraguayans, whether or not they were Marxists, had a strong notion of the "fictitious" nature of Itaipú debt.[24] The numbers were fantastically large and had to be not only translated to smaller amounts to be understood, but represented through entirely different metaphors. The juridical and financial enclosure of Itaipú fostered a sense of secrecy such that all documents and information emanating from the dam were a priori dubious. Canese and his team acted out a public nationalist drama of defense of Paraguayan sovereignty against the predations of imperialist Brazil through their interpretations of the debt, a drama that directly referenced the devastating War of the Triple Alliance (1864–70). Paper promises like those of treaties or loan agreements did not amount to much because of the commonly experienced rift between official Paraguayan government documents and on-the-ground reality.[25] Perhaps, too, the long-standing Spanish imperial relationship to performative legal writing also resulted in a popular skepticism toward promissory notes.[26]

But key details were left out of Paraguayan nationalist representations of the Itaipú debt, details that the Brazilian government and popular media were quick to identify. Central to the claims from the Brazilian government was the notion of risk. During the oil crises of the 1970s and the economic instability of the 1980s, the Brazilian government provided all the legal guarantees for the construction debt and pledged to purchase all the energy of the dam, whether or not there was sufficient demand for it. That is, the dam itself was a risky endeavor—and the debt was just an outcome of that risk. The Paraguayan Hydroelectric Commission, on the other hand, did not configure the dam itself as a risky investment. Moral virtue aggregates to the bearer of risk; in the version of the debt history narrated from within the Brazilian government, Brazil alone assumed both the risk and the virtue.[27] And, of course, 95 percent of what had been paid on the debt had come from Brazilian consumers. Brazilian minister of mines and energy Edison Lobão's willingness to state, in the midst of the fraught negotiations, that the government of Paraguay had not contributed financially in any way to the construction of the dam was an unmistakable challenge.

"After long negotiations, Brazil agreed with what Paraguay demanded. Paraguay didn't want to spend a dime—a condition set by Stroessner," said the Brazilian government official from chapter 2. "I consider [Ricardo

Canese] honest. . . . He's an idealist and I respect him, but I disagree," he continued. Of the public threat by Lugo, Canese, and the Hydroelectric Commission to bring the matter to international arbitration in order to render a decision in Paraguay's favor, he confidently said, "Let us take it to The Hague. Let us. I am so convinced of the technical, juridical [merit of Brazil's position]. . . . To see one country robbing another. No tribunal in the world would disagree."

In public and in more private venues, Brazilian government officials were confident in the legal grounding for the debt. If in Paraguay the debt exemplified imperialism where Brazil strong-armed its smaller neighbor, in Brazil, attempts to lump the debt only on the Brazilian side of the dam were yet another instance of indolent Paraguay trying to get something for free from hardworking Brazil. As negotiations wore on, the Brazilian government yielded in some significant way on most of the Paraguayan requests, including increasing the price paid to Paraguay for its energy sold to Brazil. But it did not commit itself to any action whatsoever regarding the debt. In the thirty-one-item Joint Declaration signed in July 2009 by Paraguayan president Fernando Lugo and Brazilian president Lula da Silva to resolve the diplomatic crisis, only one line referred to the debt. Item 15 stated that "President Fernando Lugo reported on the audit of the Itaipú debt being performed by the Office of the Comptroller of the Republic of Paraguay and his intention to transmit its findings to Brazil."[28] For the sake of comparison, all other items in the declaration included action from one party in response to the other except for Item 15. The Brazilian government treated the promissory notes as firm commitments, regarding the debt obligations as binding distillations of the risk. And yet risk was an unstable moving target because the dam was far less risky in 1984 than in 1974 and much less riskier still in 1994. Once energy production started, the risk of failure, which the higher interest rates were to mitigate, plummeted. The dam could have had a tiered interest approach where rates dropped with significant drops in risk, a perhaps more appropriate linking of interest rates and risk. Instead, the lack of solvency was the tipping point.

Energy technocrats from both countries, on the other hand, did not participate in the moral outrage that underscored the nationalist narratives of their compatriots. Paraguayan and Brazilian energy technocrats like Esteban Solanzo were involved in the more cordial, less media-saturated negotiation over Itaipú, paralleling the polemical engagements between the Hydroelectric Commission and Brazilian diplomatic corps. "I understand the right to obtain greater benefits, to correct errors in how the treaty has been applied," Solanzo said, speaking of the Canese-led team. "But they're using an incorrect base, and so their position is wrong." Instead of metaphors involving manioc and relatable numbers, Solanzo's story operated in US billions, in gigawatt hours, in prime rate adjustments. Whereas the Hydroelectric Commission had turned to smaller numbers and analogies to explain the finances of the dam, Solanzo regularly worked with figures at the scale of billions as part of his ordinary job responsibilities. The amounts and the differentials were meaningful to him, and he assumed that they would be for his one-person audience that afternoon, as well. For energy experts like Solanzo, it was important to rationally acknowledge the aspects of the Itaipú debt drama that fell to Brazil's favor and those that benefitted Paraguay. It proved that they were neutral and competent administrators, particularly amid popular accusations in Paraguay that Paraguayan energy elites had betrayed Paraguay's sovereignty in the dam.

Though they were out of the media spotlight, the dam's technocrats were responsible for the day-to-day functioning of Itaipú. And those who were located at the physical power plant had a tangible understanding of what binationality meant and of the ways sovereignty and energy decisions were made on a daily basis. Paraguayan electrical and civil engineers like Solanzo and Velázquez often had advanced degrees from abroad—many of them had even studied in Brazilian universities. To the engineers who spent their careers cooperating across the binational border, the story of the debt was one of pragmatism and financial exigency. Brazil offered the only viable market with sufficient demand for the electricity; looking elsewhere would simply not work for Paraguay. Because of that market's crises, the debt had grown, but changes had been made such that the debt would be fully paid off by 2023.

In spite of these claims to financial exigency, Henry Winters saw the history of the debt not so much as benign incompetence or pragmatism, but as intentional conspiracy to commit fraud. His description of the relationship between Eletrobras and Itaipú was akin to a company store in a coal-mining town where the coal company had figured out that more money could be made from the company store than even from selling coal. And in Delgado's story, the debt was a long-term strategy on the part of the Brazilian government to become the sole proprietor of the binational dam. Hearing such speculation took me aback (even if conspiracy theories are something of a national pastime in Paraguay). Winters and Delgado were world-traveled, financially savvy professionals who had tempered political analyses and were not prone to hyperbole. While on the field, I

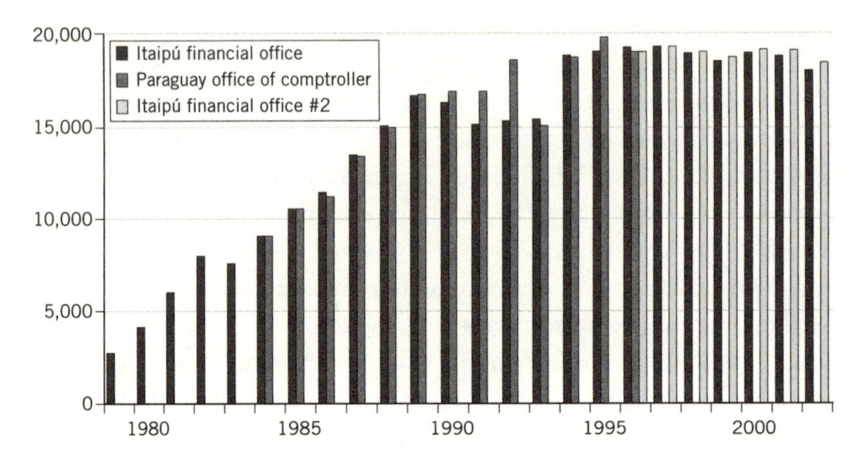

FIGURE 4.3. Discrepant reporting of Itaipú debt, US dollars.
Source: Itaipu Binacional 2003; Contraloría General de la República 2012.

took their suspicious assessments to be in spite of their professional expertise. But perhaps their suspicious assessments came from their professional expertise.

Because of the ongoing secrecy regarding Itaipú financials, debt reports from a few different authoritative sources were compiled for this chapter: *Itaipú Memorias* (Annual Reports) from 2006–17; *Prestación de los Servicios de Electricidad y Bases Financieras* (a financial compendium) published by Itaipú's Financial Office in 2003; *Segundo informe avance de auditoria* (the preliminary report on the audit) published by the Paraguayan Office of the Comptroller in 2012. Figure 4.3 shows what the Itaipú Financial Office and the Paraguayan comptroller reported as the Itaipú debt, even though the latter's numbers were supposedly based on the former's. The divergence between the official numbers presented by the Itaipú Financial Office and those from the Paraguayan comptroller raises troubling questions about everything else they (or I) have presented regarding the Itaipú debt. The difference between the numbers they report for the debt is as much as $3 billion and, even in less egregious years, they do not coincide by hundreds of millions. Neither is one consistently off in one direction—at times Itaipú reports a higher debt, at others, the Paraguayan comptroller does.

Even more confusing is that, within the very same Itaipú 2003 financial compendium, the financial office reports different quantities for the same numbers. While its general summary of debt (on page 86) to its date of publication flatly disagrees with the Paraguayan comptroller, on a chart on page 100, the financial compendium lists numbers that contradict what was printed just a few pages earlier. The few numbers listed on this second chart actually do coincide with those of the Paraguayan comptroller. If this is a problem of units, that is unclear. Neither the Itaipú Financial Office nor the Paraguayan comptroller says anything about whether nominal or current US dollars are used as the unit of measurement. If these official government financial documents about one of the most controversial topics in South American energy are so divergent, especially when it is clear that both groups have an interest in credible reporting, what is to be made of all the information they report? For all the claims to authority and authenticity, even the most basic "facts" are up for grabs.

This suggests that the numbers themselves do not form the basis of reliability, but rather that there are established narratives and points of view into which figures are inserted. Neither the dam nor the Brazilian or Paraguayan governments have seen fit to point out this discrepancy. Perhaps they think it unimportant? In Nils Bubandt's work on suspicion as empathy, a proven forgery was still taken as "real" by members of the Muslim community in Indonesia because it adhered to the view they "imagined" Christians in Indonesia to hold.[29] In the Itaipú case, interest plays a guiding role in what kinds of things are suspected and by whom. Winters and Delgado first began with the question of "What is the interest of this party in Itaipú?" and from the "empathetically" imagined answer to that question, judged and in fact constructed histories. That is, what Delgado and Winters saw as Brazil's actions, in spite of data to the contrary showing that the debt would be paid off in time, was what they thought benefitted Brazil. The same holds true for the other Itaipú debt interpretations recounted here. Lugo, Canese, and Laíno saw Paraguayan popular interests harmed by the 64 percent of the tariff price that went to debt, a situation that benefitted Eletrobras. The anonymous Brazilian government official and the more public energy minister Edison Lobão, instead, saw a continuation of previous Paraguayan practices of not contributing "even one cent" in exchange for hydroelectric riches. The mysteries of the debt allowed all kinds of narratives to be projected onto the dam, which in turn allowed the Paraguayan and Brazilian national states to perform any number of roles of villain or victim.

Discrepancies between Itaipú and Paraguayan government descriptions of when and how energy was first distributed are but one of the consequences that arose from the state-within-a-state sequestering of the dam. Not only were documents kept outside of the purview of the Paraguayan government, but discussions where financial decisions like loan agreements and fixing the base rate of electricity also took place in closed-door meeting rooms at the dam itself. This had two prominent effects in Paraguay and Brazil. The first was a mystification of money and debt. The numbers associated with the dam—billions of dollars, tens of thousands of gigawatt hours—were, by orders of magnitude, larger than the kinds of amounts most commonly used by ordinary Brazilians and Paraguayans, to say nothing about the units of measurement. Today, the average annual income in Paraguay is a little more than US$3,000, and in Brazil it is around US$7,000 a year. If US$3,000 to US$7,000 is what most people live on, then both US$1,000,000 and US$10,000,000 seem unimaginably large, and, from the vantage point of a thousand dollars, the difference between US$100,000,000 and US$1,000,000,000 becomes meaningless. If a few thousand dollars can pay for a modest living for a year, what can be bought for a million dollars or a billion dollars? The scale of the amounts made the entire financial side of the endeavor "unreal," a fiction of words and zeroes.

The second effect, on the other hand, was supremely apparent to anyone within Paraguay. Promissory notes, contracts, billing statements, checkbooks were all kept secreted within the confines of Itaipú. But the rapid enrichment of the "barons of Itaipú" (*barones de Itaipú*), those fortunate enough to work for the dam or to acquire one of the lucrative construction contracts, was evident. Estates, mansions, luxury cars, designer clothing, and jewelry were patently obvious even if the alleged embezzlement, kickbacks, and overcharging were hidden in paperwork squirreled away behind concrete walls. Employment within Itaipú became synonymous with wealth. Paraguayan salaries were set at parity with Brazilian salaries in the dam, even though the cost of living differed. And the director positions—with on-the-books salaries above $240,000 a year—inspired even more suspicion. Two of the "barons" wrote important texts, replete with photocopies of official Itaipú documents before they were readily available to the Paraguayan public: Enzo Debernardi (1996), the first Paraguayan executive director of the dam, and Juan Carlos Wasmosy (2008), a lead engineer during construction and erstwhile president of the Republic of Paraguay during the debt renegotiation of the 1990s.

Gustav Peebles's *The Euro and Its Rivals* describes how a Swedish-Danish binational bureaucracy managed unwieldy assets and obligations in order to foster a new citizenship identity.[30] He found that proponents of a binational/regional identity encountered challenges from subnational and national loyalties, specifically expressed through disagreements over monetary policy and what currency should be used. In the Itaipú debt case, competing national and professional identities surfaced as disagreements over what makes debts and debtors good or bad and what counts as a good enough reason to not pay back a debt. Economic anthropologists point out how relational proximity (e.g., kinship) affects loans and repayment expectations. Zero or negative interest rate loans may be expected between family and friends but count as theft or bankruptcy between more distant relations. Itaipú is fraught with ambivalent relations between the two state owners and between the dam and its creditors. To Paraguayan advocates of loan forgiveness, obligations were framed as a result of the negative moral character of loan signatories—spurious and usurious debts caused by imperialism and dictatorial puppets. To Brazilian audiences and energy technocrats, the debt obligations were binding and the desire to slip out of them a demonstration of Paraguayan moral deficiency.

But the binationality of Itaipú, an outcome of the geographic location of the hydroelectric resource, complicated the relationships even further. If economic anthropologists have demonstrated that kinship influences interest rates, then what kind of interest rate is expected in ambivalent citizenships? Since citizenship is a relationship of obligation between state and nation and has become the most common context for making claims on the state, binationality allows for a situation where one national community may make claims of a different state. During Lugo's regime, the Paraguayan national community sought loan forgiveness from the Brazilian state in Eletrobras. No major contemporary Brazilian group advocated for Brazilian state loan forgiveness of the Itaipú debt for Brazilian consumers. Common sense dictates that it is more appropriate for states to forgive the debts of their national communities before those of a nearby foreign one. The binationality of Itaipú, which made the Paraguayan request possible, made loan forgiveness politically impossible. How could Eletrobras tell Brazilian consumers that they had to pay a debt that the other half the dam did not? Without outside pressure, it was untenable.

Conclusion

A team of analysts from Columbia's Earth Institute headed by Jeffrey Sachs offered their own interpretation of the state of Itaipú finances, asserting that the interest rates on the Itaipú debt were too high in the 1980s and that the tariff should have been higher from the start, at least enough to cover the minimum. Instead, they claimed that an industry standard "real interest rate" of 5 percent and a "real price" for exported energy would have been more appropriate.[31] The institute found that interest payments of over 5 and even 10 percent per year "seem[ed] too high, especially as [the dam] can be considered a low risk investment since it was entirely collateralized by its own revenue stream."[32] "Based on these assumptions," wrote the Earth Institute research team, "and allowing for reported operating and maintenance costs, we find that Paraguay has in effect paid off its portion of the IB [Itaipú Binational] debt by 2013."[33] The ebullient Paraguayan media widely broadcast the Earth Institute's findings.

Predictably, within Brazil, the reaction differed. Rubens Barbosa, former ambassador to the United Kingdom and the United States and then president of the international trade wing of FIESP (the largest professional association of Brazilian industry) wrote an op-ed baldly declaring that "the aim of Paraguay" was "without having invested, to own half of Itaipú."[34]

The Earth Institute's findings reproduced conceptual representations common to the Lugo government approach. Unlike official documents and speeches from Itaipú personnel, the Earth Institute text did not treat Itaipú as a unitary whole, a legal entity separate from both the Paraguayan and Brazilian states. Nevertheless, according to law and accounting practices, the dam's finances are technically undivided masses of dollars—signaling the undifferentiated nature—within dam, though separated in Brazilian and Paraguayan halves in common parlance. Thus, the institute's suggestions imply a very different organizational structure within Itaipú. In any case, it is doubtful whether any report anywhere could easily dislodge the Brazilian position on the legitimacy of the Itaipú debt. The usefulness of Sachs's interpretation lay not so much in that it could convince the Brazilian government, but that it gave legitimacy to Paraguayan complaints on an international plane. The Paraguayan government successfully elevated the problem of the Itaipú debt to an IMF investigatory committee in 2013.

Creative financial regimes and complicated accounting schemes arise in moments of financial constraint (or new opportunities), often accompanying a scale jumping that uses a spatial fix to resolve politic economic crises.[35] The path of licit money into and out of the dam highlights the curious equivalence between three different accumulation logics: (1) rent, particularly the low, but guaranteed, amount the Paraguayan government received for energy exports, (2) rather than a risky but potentially more lucrative market price and, finally, (3) compound interest that accrued only to the Brazilian side. Itaipú's financial regime reinforced sovereignty and security as important values and, in fact, defined them. For all the claims to the contrary, neither was the invisible hand the main arbiter of Itaipú energy finances, nor was the treaty sacrosanct. Instead, the finances lay at the mercy of constant negotiation by technocrats as they generated currency out of water. And from the very start of the binational project, the Executive Directorate and the Governing Council both offered flexible applications of the treaty; "inviolability" and "respect" for the agreement served as rhetorical markers to bolster a position statement rather than as a description of previous events. Itaipú, the very engine of Brazil's spectacular industrialization from the mid-1980s to the present, possesses decidedly nonmarket characteristics. This is not just the resilient persistence of precapitalist mechanisms; nor it is merely about the incorporation of these things into modern economic practice. Rather, the finances of the world's largest dam show that the intentional creation of nonliberal practices is the very sustaining core of the liberal market economy. Moreover, the now partially privatized monopoly guarantees returns to its investors.

The history of Itaipú suggests that the materiality of hydroelectric renewable energy (the fixity of sources, the uneven distribution of demand, the technical infrastructure requirements) enables novel political-economic formations even as the state plays a more traditional, ongoing role in maintaining the free market economy. As the Paraná runs through multiple countries, the dam became an interstitial space in global finances, where transactions not sanctioned within normal governance were now possible. Itaipú took on qualities similar to the Cayman Islands or the free trade zone of Ciudad del Este just eight miles to the south. Beginning in 1970, physical warehouses and then, later, specific documents and stamps created pockets within the geographic territory of Ciudad del Este, Paraguay, that assigned a privileged tax regime in order to foster the "reexport" trade, where goods were brought cheaply to Paraguay for resale to Argentine and Brazilian consumers. Complex and innovative financial regimens that took advantage of borders and liminality were a trend of the 1970s. However, whereas the Cayman Islands or Ciudad del Este benefited corporations or private individuals, the exception fostered around Itaipú intentionally profited the Brazilian and Paraguayan states, actors, and institutions.

But the financial creativity in Itaipú goes beyond the incorporation of illiberal mechanisms or the Caribbean tax haven–like exceptionality. And this is where the anthropocenic moment comes in. The dam's financial complexities function as Anthropocene-compensating strategies, facilitating continued access to a sustainable energy resource that can thus support industry in a time of fossil fuel scarcity just as Smith-Nonini claimed regarding neoliberalism's connection to oil insecurity and debt. Renewable energy, superintended by the state, makes it clear that state institutions play an ever-present leading role in market formation. Although not coinage per se, the debt origins of electricity suggest money-like characteristics. Itaipú hydroelectricity has nationality and can store political value. Through the dam, like a currency exchange, electricity can be converted into US dollars, Brazilian reales, Paraguayan guaranies. And, perhaps in a fully robust regional state apparatus of the future that governs human and natural resources (especially in light of severe climate stresses), electricity is imaginable as a postmoney currency itself, circulated and hoarded by the state. In the Anthropocene, major renewable energy resources may become complex financial institutions and, as a dam takes on the qualities of an investment bank, electricity goes from being a current to becoming currency.

5

Neoextractivist Futures

Introduction

FEBRUARY 2009

Months before the Joint Declaration was signed, Carlos Mateo Balmelli explained his view of development.

"I was reading Arendt's *On Revolution*," said the executive director as we sat in his office. "That is her best book. She quoted Lenin, who asks how you will solve the—in Spanish, we say 'cuestión social.'"

"That's how we say it in English, too: the 'social question,'" I said.

"How will you solve the social question?" he continued. " 'Electrificando' [electrification] was Lenin's answer. Have you seen *Doctor Zhivago*?"

"No."

"You need to correct that. Omar Sharif leaves a big energy plant." Visual shots of a massive hydroelectric dam open and close the 1965 Academy Award–nominated film starring Omar Sharif. But Balmelli was not interested in pursuing a pop-cultural tangent.

In the very next breath, he continued with the lessons learned from Hannah Arendt's comparison of the American and French Revolutions,[1] saying, "Lenin electrified Russia. This means my ideas are right. Russia was a traditional society. With this decision, the revolution, they industrialized. [Berkeley professor] Bender says the Russian Revolution was a way to modernize a traditional, rural society. There are different paths to modernize:

the French model, the German model. . . . The Russian Revolution was not a socialist, but a modernizing one. So, electrification is important."

Paraguay, the comparison implied, was also a traditional society in need of modernization and industrialization. Balmelli's political and ideological leanings held no affinity for the socialist transformations of the early twentieth century, but he was interested in how a largely agricultural society so quickly converted into an industrial one.[2] He concluded with a broad-strokes explanation of the Paraguayan economy and the potential for the country. "There are three important economic sectors in Paraguay," he said. "(1) The *ganadero* [ranching]. But it doesn't have the effect to include the people. It's *excluyente* [exclusive]—not the trickle-down effect. (2) The *agricultura mecanizada* [industrial agriculture] is like the first. (3) Electrification includes people, giving the people electricity to use as a tool."

Balmelli saw electrification as the only way out of Paraguay's underdevelopment. Of the three major economic sectors in Paraguay, the first two—ranching and agribusiness—were exclusive and would not impact the people because neoliberal promises of trickle-down effects had not worked. But because electrification reached individual households and drew people into industry, it was a tool for modernization that could end poverty. That is, electrification via a renewable, sustainable energy resource would lead to the industrialization necessary for sustainable growth. In Paraguay (and beyond) discussions about how to use Itaipú for hydroelectric-led development participated in an urgent theoretical debate regarding the value of natural resources and the place of extraction in the twenty-first century. Increased government attention and the malleability of a new regime were only part of the urgency. Determining how to use electricity and revenue took on added importance because of the rapidly narrowing time horizon of the 2023 expiration of the Itaipú Treaty, when all price and energy sales agreements would lapse.

The signing of the Joint Declaration in July 2009 put new economic possibilities into play. The two Paraguayan electricity sector teams differed in more than just their postures in initial negotiations; they presented two distinct approaches for how to wield Itaipú in development as well. This chapter performs a kind of postmortem as we analyze models of hydroelectric-led development, how policy unfolds, and why politics as usual took over, demonstrating fragilities attendant to hydrostates. To understand how energy policy gets worked out, I will trace two roads not taken, two visionary stances on Itaipú and political-economic futures in light of the Joint Declaration's opening and what happened instead. The natural experiment from chapter 3 continued as regional energy experts considered an Itaipú technocratic model of industry-led development through public-private partnerships or a model of government-led development through targeted job creation and financial investment proposed by the leftist activists in the Commission on Hydroelectric Binational Entities (CEBH). Yet in spite of the limited window of opportunity signaled by the 2023 deadline, the traditional model of rent redistributed to government coffers for recurring costs, electricity destined for increasing residential consumption, and institutions oriented toward parochial internal-to-Paraguay political economic dynamics won the day.

Itaipú was a way of speaking, planning, and funding the future. If Itaipú in the Paraguayan imaginary brought wealth and the creation of a middle class in the past, in the Lugo present it rendered two tangible offerings: money and electricity. In the first year of Lugo's presidency, 2008, Itaipú paid $2.064 billion ($1.032 billion for each side) toward its construction debt and $240 million to the treasuries of each country in the form of royalties. Each executive director had as his responsibility an annual budget of $307 million out of which to pay for salaries, operations, maintenance, and social programs. Paraguay's treasury also received $117 million in compensation for its excess energy ceded to Brazil. And once the Joint Declaration was ratified, an additional $240 million (for a total of more than $600 million) would be injected into the Paraguayan treasury.

How to administer these royalties and ceded energy payments (a decision made by the executive branch) and how to administer the Itaipú budget for social programs (a decision made by the executive director) lay at the center of expectations of what the Paraguayan state ought to do for the Paraguayan nation. Under the Colorado administrations prior to Lugo, these were parceled out according to personalist logics and party loyalty. But with the election of Lugo, this, too, was up for grabs. In a country where the government's budget was $6 billion in 2008 and the GDP was $16 billion that same year, these amounts had the real capacity to effect a dramatic shift—a financial order of magnitude change that would be felt almost immediately. Itaipú had transformed the organizational logic of the Paraguayan state in the 1970s; by reengineering it again in the 2000s, new political actors in Lugo's government sought to attain the development futures they thought most prosperous and desirable.

In earlier chapters, I made an argument to consider Itaipú as a source of multiple circulations, from electricity to energy rent to debt, each with constraints both political and material; in this chapter I explore how hydroelectric-led development in the twenty-first century reimagined hydrodollars. Hydrodollars may be thought of as economic growth and industry and commerce powered by hydroelectricity, government liquidity through energy rent capture, and increased capital circulation within a country because of the spending of rent and wages. Hydrodollar dynamics dominate within Paraguay because industry and commerce are supplied by hydroelectricity; fossil fuels are imported for use in transportation and cooking.[3] I have insisted on arguing that the energy source affects the political-economic formations possible. This means that as anthropogenic climate change impacts solar, wind, tidal, and hydropower, the economic growth fueled by these sources must also adjust. Economies and markets, not just rent-reliant government agencies, are encumbered by the kind of power that powers them because although electricity is fungible, energy sources are not.

In many ways, the triumph of July 2009 was just the start of a long process. While the presidential signatures moved the changes forward, congressional approval from both countries would be needed before formally implementing the declaration. Much wooing remained to be done. Paraguay's midterm elections were slated for 2010, and, with the signed agreement in hand, the Lugo government finally had something concrete to offer constituents. From the very beginning of Lugo's tenure, the pledge to invest Itaipú's bounty in the nation as opposed to previous administrations' cronyism distilled the *promise* of the nation. On the one hand, the new administration of previously marginalized groups vowed to use the state apparatus to serve the nation—"promise" in the sense of commitment from the state to the nation. On the other, "promise" also describes potential for the future, as when adults say, "that child has great promise." With the vast Itaipú resources now in new hands, there was a sense that Paraguay had promise, that the country's future could be prosperous, equitable, different from its past. Although earlier chapters feature perhaps more obvious conversions, from energy into money into debt, the dam also afforded a way for energy managers to convert Paraguay's political and economic machinery.

For the Lugo administration, bolstered by the success in bringing Brazil to the negotiating table and attaining valuable concessions, the central question regarding Itaipú was a broader one about the fundamental rules of economic growth. To many in Paraguay, Itaipú served as a synecdoche for the Paraguayan state because of the country-shaping impacts of hydrodollars. Disagreement over the forms of government and the forms of economy proposed by the leftist activists in the Hydroelectric Commission and the more traditional, liberal technocrats in the Itaipú Executive Directorate extended the competition between political-economic strategies from chapter 3's negotiations. Though rent was the primary accumulation strategy within Paraguay for the first thirty-five years of the dam's existence, both energy teams sought to dislodge this and foster new postures toward the resource potential of the Paraná River. In advocating energy extraction–led growth, they combined political ecology with political economy, drawing on assumptions about the very nature of nature.

The proposals advanced by the Hydroelectric Commission and the Itaipú Directorate were part of a larger, regional reaction against the failed promises of neoliberal reform and export-led growth. Skyrocketing soy production had increased Paraguay's GDP by 200 percent from 1998 to 2008, but the rising tide had not lifted all boats for several reasons.[4] As Balmelli pointed out, industrial agriculture, because it was mechanized, created relatively few jobs. And the low tax burden in Paraguay (in part because the government could rely on energy rent for revenue) meant that economic growth from soy was not reinvested in Paraguay but was rather retained by large-holder farmers or transnational companies. Paraguay's 2008 GNI per capita was $2,366, and 37.9 percent of the population lived below the poverty line in 2008, an increase from 36.1 percent below the poverty line in 1998 even as soy wealth flooded the country.[5] Soy was just the latest in a string of primary product exports sometimes controlled by local interests, often developed via foreign capital, that fostered economic enclaves and few backward linkages—a familiar story of economic growth in Latin America.

With fiery rhetoric, mass social movement mobilizations, and denunciations of globalization's uneven development, progressive governments elected in the early 2000s came to power in South America with a popular mandate to address ills stemming from neoliberalism. Rather than choosing either a market-rejecting model or continuing the neoliberal experiment, they tried to find a new way to use energy and mineral exports to expand benefits beyond elites. New leftist governments put into practice political economic regimes that have been called "progressive neo-extractivism" and "neodevelopmentalist extractivism."[6] Across the continent, neodevelopmentalist extractivist policies were characterized by strong state involvement, expanded social investment in education and other services, and in some cases even direct cash transfer programs like the Bolsa Família in Brazil and the student-attendance-based Juancito Pinto subsidy in Bolivia.[7] By the time both Paraguayan energy teams laid their proposals for Itaipú and growth, they had the examples of neoextractivist experiments from Argentina to Venezuela to consider for hindsight. Analysts of progressive neoextractivism have pointed out that these approaches were still rooted in a hope that export-oriented development could, if properly steered, bring social justice. And the dominant view of nature was Western: nature envisioned as a source of resources primarily to be used for the benefit of humanity. For some critics, the philosophical underpinnings behind any form of extractivism, progressive or neoliberal, rendered a fatal flaw to the system. But for others, the problem with these models is the emphasis on the export of primary products rather than on industrialization.

This chapter explores what can be learned about twenty-first-century development models from the policies proposed by the two teams. Importantly, the two policy proposals intentionally counteracted economic and political hazards associated with hydrodollar dependency, the resource curse, and underdevelopment. The Joint Declaration's thirty-one items centered on three general areas of bilateral concern: electricity, Brazilian and Paraguayan citizens residing in the other country, and commerce. In this chapter we will see how the Hydroelectric Commission and the Itaipú technocrats interpreted the development potential offered by the declaration. Although much of the document developed in the Paraguayan side Itaipú, it bore marks of compromise with the Brazilian government (such as a tripled, not quintupled amount for ceded energy) and with the Hydroelectric Commission/Paraguayan Foreign Ministry (such as Item 15's mention of intergovernmental communication regarding the debt). As table 5.1 shows, the main thrust of the agreement was to raise the amount paid for cession of energy, to affirm the right of direct sales of Paraguayan electricity in Brazil, and to commence infrastructural projects. One notable feature was that none of the immediate financing for any of the changes came from the Paraguayan government—funding would come from Brazilian consumers, debt held by Itaipú, or financing via the Brazilian Development Bank. As importantly, the solution to the problem of underdevelopment was deeper integration.

But for all the enthusiasm for the Joint Declaration in Lugo's government, the seeds of Paraguayan partisanship reached their toxic maturity in the months after July 2009, hobbling the implementation of the agreement. Of all the ambitious, hard-fought energy compromises in the Joint Declaration, only one ever made it to law: only the increased compensation, the lowest-hanging energy fruit, was drafted into an amendment (*nota reversal*) to be debated and voted on by either legislature. Though the Paraguayan

TABLE 5.1. Summarizing the Brazilian-Paraguayan Joint Declaration of July 25, 2009

- Six Points affirmed; each point addressed with pledges from Brazil for change in the direction proposed by the memorandum except Point 2 (renouncing the "spurious debt" of Itaipú). (Items 13, 14, 15)
- Multiplication factor raised from 5.1 to 15.3, changing amount for ceded energy from about $120 million a year to about $360 million; amount received by both countries for royalties also increased slightly. (Item 5)
- Direct sales to Brazilian market via ANDE permitted as soon as possible; possible Itaipú electricity sales to 3rd countries following 2023. (Items 6, 7, 8)
- Energy infrastructure in Paraguay funded and built by Itaipú to increase Paraguayan consumption (including 500 kilovolt high-tension line) at no cost to Paraguay. (Items 4, 9, 10)
- Unified Tariff Regime (RTU) and other Mercosur-related financial mechanisms approved, to the benefit of merchants in Ciudad del Este, Paraguay. (Items 18, 19, 20, 30)
- International bridges, development projects, potential railway linking Paraguay to a Brazilian port built/funded by Itaipú or BNDES (the Brazilian Development Bank). (Items 11, 12, 16, 17, 21, 22, 23, 24, 26, 27, 29)
- Support expressed for Brazilian/Paraguayan citizens living in the other country. (Item 25)

congress passed the amendment formally constituting the Joint Declaration by early November 2009, the Brazilian time line dragged on for years. Only in May 2011, months after Lula's successor took office, was the amendment presented to Brazil's congress and eventually, though begrudgingly, passed. A working group to finally discuss how to implement the financial interventions (namely, increased compensation) convened in October 2011. Instead of either of the robust policies proposed by the energy teams in 2009, the additional resources were directed eventually in 2012 to a newly formed Public Investment and Development Fund (FONACIDE), which followed in the tradition of elite rent capture and nonproductive capital investments. At best, FONACIDE envisioned Itaipú revenue as support for education, health, and direct government expenses; at worst, it provided assets to be distributed patrimonially across patron-client relationships.

Strategies to reform government embed critiques of what was wrong with what had come before; Paraguay's underdevelopment was cast as a result of bad government, the mismanagement of the past. The contest within the Paraguayan government over how to invest energy in progress shows how policy gets debated and implemented. Unexamined within both proposals was the expectation that extraction could and should be converted into positive financial futures for Paraguay. Here, I argue, we will see what characterizes two New Democrat Paraguayan narratives—one liberal democratic and one leftist—about development, good government, and the ideal future that were in competition to attain hegemonic control over the way hydropolitics was thought and practiced. And as we do so, we will answer broader questions about political transformations in Latin America's leftward turns, about how new narratives of futures are fashioned out of older ones, and about the concrete political and economic solutions that arise from ideologies bundled into left-turn politics. But this chapter is also a chronicle of a defeat foretold because rent was the path of least resistance.

CEBH: Participatory Democracy, Government-Led Investment, and Regional Energy Integration

Once the Joint Declaration was signed (July 2009), the Hydroelectric Commission continued its well-established pattern of holding open meetings and listening to feedback from those who attended, but the discussions changed to what to do with the earnings (*beneficios*), the increased monies Paraguay would gain from Itaipú ceded electricity and royalties. As we will see, for the Hydroelectric Commission, a major problematic result of Itaipú energy rent capture and simple redistribution was the development of a hydrostate resource curse. And so they focused their work on tackling Dutch Disease[8] and public expectations of perpetual energy rent, two prominent hydrodollar effects that resulted from the political economy of Itaipú rent and financial surplus.

The high-water mark for the Hydroelectric Commission's development planning was a policy wonkish three-day seminar held at one of Asunción's most exclusive hotels from December 12 to December 15, 2009, which explored three themes: (1) the environmental impact of Itaipú and Yacyretá, (2) the use of the additional proceeds from Itaipú, and (3) international energy integration. Though not in its name, the purpose of the "Working Group Seminar: Utilizing the Extraordinary Earnings Proceeding from the Binationals" opening event was to build consensus around a law being proposed by the Hydroelectric Commission that would create a "Fund for the Promotion of Economic and Social Development" (not the fund that was passed in 2012). The prospective law was the Hydroelectric Commission's recipe for development, a policy that would determine how additional Itaipú revenue would be invested in infrastructure, economic development, education, and long-term financial planning (see table 5.2). It called for the creation of a permanent fund under the Ministry of the Treasury, which presenters spent the day carefully explaining.

TABLE 5.2. From the Law for the Fund for the Promotion of Economic and Social Development

a) 25% on infrastructure projects for the transmission and distribution of electricity;
b) 20% on infrastructure projects in general, with an emphasis on infrastructure for potable and sanitary water; rural roads and infrastructure for agrarian reform; infrastructure for education and health; infrastructure for scientific and technological research;
c) 5% as collateral for international loans;
d) 25% for the Fund for the Development of the Productive Sector and the Environment, with an emphasis on small and medium-sized producers, particularly for impoverished sectors; for the construction of lower-income housing, with a preference for mechanisms of self-administration. In the case of productive projects, those that generate the most employment of highest quality will be prioritized;
e) 5% for the Fund for Education and Scholarships;
f) 10% to subsidize the electricity tariff and a pension for the elderly;
g) 5% for the Anticyclic Fund for Economic and Social Stability;
h) 5% for the Emergency Fund, for catastrophes and epidemics.

Financial mechanisms provided by the law were designed as antidotes to specific economic illnesses. Nearly half of the additional revenue was destined to construction projects emphasizing electricity distribution, water, and roads. One quarter of the fund was designated for business development under the rubric of "Development of the Productive Sector and the Environment," but even under that, there was an explicit provision for the construction of low-income housing. Presumably, job creation would occur as a part of the construction boom anticipated by the bill. Both the content and the form of the law proposed by the Hydroelectric Commission signaled a discrete understanding of how economic development would come about. And the contrast with the plans of the Itaipú technocrats was notable, even in just the fact that the very proposal for guiding economic growth took the form of a law to be passed by congress and not of, say, a contract between firms.

More than one hundred men and a few women from around the country—the vice minister of mines and energy at the time was also a woman, Mercedes Canese (a trained engineer, like her father, Ricardo)—plus a handful of international visitors participated in the Saturday seminar (timed to not compete with workweek responsibilities). The morning panel discussions featured Paraguayan government experts, Hydroelectric Commission members, and even a former Bolivian minister, each followed by lengthy questions and answers from the audience. A carefully curated series of presenters demonstrated why a bill mitigating against Dutch Disease was deemed necessary by the Hydroelectric Commission, illustrating the economic philosophy behind its proposals, and hinting as to why it was never passed and instead replaced by an edict that destined energy rent to short-term costs.

To open the morning and lay the foundation for the other days, the first panel focused on background and context. A treasury technocrat began a lengthy PowerPoint presentation, cautioning against risks facing Paraguay as a result of the anticipated increase in income. He repeatedly urged against local governments using the revenue to pay general expenditures—salaries, recurring expenses for infrastructure, unproductive outlays of capital, and so on. In doing so, the economist raised the same troubling statistic in public as Hugo Mesa had in private: in parts of the interior of the country (i.e., not Asunción), royalties from the dams provided up to 40 percent of the income for municipalities. And because local governments had grown accustomed to depending on royalties to meet their expenses, there was a danger that they would get into greater debt and an eventual systematic default as they overanticipated a windfall following the Joint Declaration.

The peril, he warned, was "enfermedad holandesa" (Dutch Disease)—that with the increased exploitation of one natural resource, Paraguay would become increasingly more dependent on the proceeds of these sales to the detriment of other economic sectors. Because the additional revenue had a limited time span, if Paraguay did not take advantage of the funds in strategic ways, the country would actually be worse off than before. Once 2023 arrived and the 1973 treaty financial agreements expired, everything from baseline prices for Paraguay's ceded energy to royalty amounts would no longer be valid. Using an expert lexicon of "unproductive capital," the treasury representative admonished the audience against using the additional monies for recurring expenses, painting a dire picture of what would happen after 2023.

If, he said, Paraguay destined Eletrobras compensation and any revenue from ANDE exports to ordinary government expenses, rather than building up the capacity to absorb the dam's energy, the country would find itself atop of a vast energy surplus. Lacking any preset agreement Paraguay could find itself forced to sell at a disadvantageous price. This, in the middle of technical jargon, made a visible impression on the audience. Instead, the economist urged the creation of several natural resource funds—an Anticyclic Fund (for economic crises) and an Emergency Fund (for natural disasters). Chile's Economic and Social Stabilization Fund (ESSF, formerly the

Copper Fund) served as the explicit model for this scheme. The Andean country had invested additional revenue from bountiful mineral sales years into an anticyclic fund, providing a buffer that could be used in times of crisis to inject capital into the economy to prevent a slowdown in spending and thus jumpstart the economy—as they had done in 2009.

Ricardo Canese, as head of the Hydroelectric Commission, spoke all three days. For his first order of business on Saturday, he presented findings from surveys conducted by the Hydroelectric Commission during their nationwide dialogue circuit. He displayed telling answers to the question, "On what should the royalties and compensation be used?" Out of hundreds surveyed, 0 percent said, "only on recurring expenses"; 22 percent replied, "only on capital investments." However, a full 77 percent said that the monies should be spent "on both capital investments and recurring expenses" (with 2 percent not responding).[9] And so, the majority of the Paraguayans polled by the Hydroelectric Commission disagreed with the treasury economist's stand against using finite resources to cover recurring costs. This 77 percent served as the subtext for the day's presentations as not just the treasury technocrat but Canese and all presenters save one reiterated the danger of Dutch Disease and the need to invest the additional royalties and compensation solely in development funds. To that end, Canese explained the Hydroelectric Commission's proposal. Because they knew public perception, the commission crafted a popular strategy to explain why the current practice of depending on royalties and compensation for what would otherwise come from a tax base would be disastrous in the long term for Paraguay.

During the question-and-answer session, members of the audience challenged the treasury technocrat on how he could dare call plazas and public football pitches "unproductive" uses of capital—an instance where technical terminology led to misunderstanding. The audience responses served as a micro-indication of perspectives across the country. Popular expectations within Paraguay understood "renewable" to mean unending, and since the price of electricity climbed, so too should the proceeds of the dam. Perhaps this was because, on average, Paraguayans with whom I spoke were not aware that Itaipú had a finite life span, that though it was renewable, it was not perpetual. Both Canese and the treasury representative focused on the impending 2023 inflection point and not the more distant permanent endpoint of the dam's obsolescence. But no one questioned the advice against increasing hydroelectric-based revenue support to current municipal expenses or the plan to create renewable funds until a representative from the official Paraguayan Organization of Mayors and Municipal Governments (OPACI) took the stage.

"We do not base our [presentation] on 'statistics' or 'analysis' or 'graphs' or 'scientific rigor,' but rather pragmatism, real facts, and conversations with mayors," he said. The OPACI representative proceeded to enumerate the needs of local governments that should be met by the additional income: hiring and training staff, improving technology and equipment, encouraging citizenship participation, and, listed last, unelaborated "productive" capital investments. Though he was in the minority at the seminar, the OPACI official represented the perspectives and priorities of much of the Paraguayan public. In other words, like the 77 percent of the Hydroelectric Commission's respondents, he advocated a mix of undefined capital investments and recurring costs with an emphasis on the latter, never mentioning Dutch Disease.

He was followed by Álvaro Ríos Roca, Bolivia's former minister of mines and hydrocarbons, who was slated to speak on their national experience "of consensus" in deciding what to do with additional income from natural gas. "I come here without ideology, just with facts," he said to open his presentation. As an opponent of Bolivian president Evo Morales, he critiqued the social spending in the head-of-family bonus, referring to it as "vote buying." This set him at odds with audience members sympathetic toward a Chávez-Morales Bolivarian Revolution, who later challenged him during the question period.

"Brazil acts as an empire; I'll show you: Not with the military, but with negotiations," the former minister said. "They may do to Paraguay what they did to Bolivia." Then Ríos Roca explained that once Morales demanded national control of Bolivian natural gas (which had been controlled and processed by Brazilian companies) and having gained that, celebrated an anticipated boon in revenue, Brazil entered negotiations with Peru to develop that country's natural gas resources. This would mean that in just a few years, Brazil could turn to Peru instead of Bolivia, and then Bolivia would find itself without its main market, forcing it to lower its price on gas or else not sell.

"I wonder what Brazil has planned with all the hydroelectric plants it's constructing in the Amazon," he added provocatively. The intimation was that once Paraguay controlled its half of Itaipú's energy fully, it would find that the main energy market in South America no longer needed it and thus the price of energy would plummet. Though he did not use the technical phrase, he clearly saw Brazil as wielding monopsony (lone buyer) power in the international energy sector and that this would be a problem in the future. Monopsonies (and monopolies) may be the most efficient way to organize electricity services and other public goods, but in the Bolivian and Paraguayan cases, competing national interests were at play, making price agreements more challenging.[10] The problem of monopsony is exacerbated by the fixed infrastructural requirements of both hydroelectricity and natural gas, especially for the two landlocked countries: immovable high-tension power lines and natural gas pipelines connect supply to demand. If the one buyer at the end of the wires or pipes does not wish to purchase the goods, it is not possible to easily redirect the commodity elsewhere in order to find a better price. Instead, entirely new infrastructure must be constructed. And so, both Bolivia and Paraguay might find themselves with an energy surplus on their hands and a buyer demanding a lower price because of other options, but with no simple way to find another market.

We spent the afternoon in smaller breakout groups assigned to debating the tentative solutions offered by various panelists and to answer the question "how to generate consensus." My group, which looked like the others scattered throughout the ballroom, consisted of more than a dozen men sitting in a circle—urban and rural Paraguayan workers from across the country—and one Italian journalist; I was the only woman. They reiterated what had been said many times already: Itaipú money should be invested in job creation, education, and health care. Then the OPACI representative joined our circle and talked again about municipal government needs. Everyone listened in silence but disagreed after letting him speak. The teaching point of "additional monies should go to a dedicated development fund, not on recurring expenses" had taken root. Instead, they talked about how to use a natural resource development fund to create jobs in the city and in the country and in this way diminish crime and violence and avoid the "favelaization of Paraguay."

With a four-page draft of the goals and evaluation criteria for the Fund for the Promotion of Economic and Social Development in hand, our group wrote ideas on a large newsprint sheet to give back to the organizers of the seminar. While "productive capital investments" and "anticyclic precautions" were not expressions used by the larger audience, the notion of short-term, strategic investments for long-term results was often repeated and the basis for calls for education and job creation. And as for how to generate consensus, the concrete suggestion offered by the group was to have meetings like this throughout the country. The Hydroelectric Commission's Saturday workshop was participatory lawmaking in action. Through paying more than mere lip service to hearing from the people, but holding public meetings and anonymous surveys, the Hydroelectric Commission was able to craft a tailored response to engage the broader Paraguayan public.

While much rhetorical import might be laid on transparency in the new government, the Hydroelectric Commission's communications strategy was a concrete example where perhaps an overwhelming amount of data and information was made accessible to many. The Hydroelectric Commission attempted to frame the debate around Itaipú within Paraguay by hosting nearly 150 public meetings throughout the country for the negotiations in 2009 and by initiating another round for how to use the additional revenue. Open conversation did not mean, however, that all issues were equally discussed. The warning by the Bolivian former minister about Brazil's circumvention of Bolivia's gas via Peru (and potentially Paraguay through its to-be-built internal dams) raised no discussion that I could hear. And at no time in any presentation did I hear anyone mention that sedimentation over the years could fill the reservoir and cavitation would wear down the turbines rendering the dam unusable.

If day one of the Hydroelectric Commission–sponsored seminar exemplified participatory democracy and the economic development agenda of Lugo's leftist allies, days two and three elucidated the energy-economic agenda more fully and, specifically, how regional energy integration would work to bring national development. Building consensus takes time, especially across professional and educational differences, as was obvious in the Saturday debate. The Monday–Tuesday "International Seminar: Energy Integration" was attended by a different, smaller crowd—domestic and foreign energy experts, not the cross section of Paraguayan society that chose to attend the Saturday meeting. Though definitional misunderstandings did not take up time, consensus on the best plan was not necessarily easier to achieve. The two days on energy integration focused on how more electricity income might be brought to the Paraguay government, to then be administered per the Economic and Social Development Fund.

"And so, to summarize," said Canese from the front of the room on the last morning, "if we consider Itaipú a 'border plant' [*central de frontera*], it creates opportunities for earnings for Paraguay, Bolivia, Argentina, Chile, and Uruguay. Argentina, Chile, and Uruguay generate eighty million megawatt hours per year with hydrocarbons. They can absorb nine million megawatt hours per year [from Paraguay's half of Itaipú and from Acaray Dam] as substitute. The global benefit would be US$968 million per year [in savings and revenue]. Paraguay can earn . . . millions a year for exporting only five million megawatt hours without losing its compensation. This export affects barely 15 percent of the energy that Paraguay exports to Brazil (thirty-eight million megawatt hours per year). It's a solution where everyone wins, including the planet—it avoids burning hydrocarbons."

Canese's presentation conveyed technical information about energy consumption, not public opinion, and emphasized the non-Brazilian energy market. The goal was for Paraguay to sell competitively priced electricity to neighbors (since the megawatt hours produced by Itaipú were cheaper than electricity generated from fossil-fuel-burning plants) via energy "swaps" to Chile and Uruguay through the high-tension lines at the Argentine-Paraguayan dam, Yacyretá. The "global" benefit figure summed Chile, Uruguay, and Argentina's savings with Paraguayan revenue. Part of the deepened integration was based on an attempt to create the new conceptual and perhaps juridical category "border plant," moving Itaipú from the binational to a regional scale.

The presentation was so convincing that one Paraguayan asked, "Here it is obvious that there are benefits for both vendors and buyers. What are the principal political obstacles—not juridical ones? If it's so beneficial, why isn't this being done?" Canese answered, "We are living in moments of change. This will also be a process. The difficulty is in the 'national security' mindset. . . . When we asked in 1973 why we couldn't sell to third parties, the answer from Stroessner's representatives here was that Brazil would never trust in an Argentine hydroelectric plant. . . . But then in the 1990s came the Treaty of Asunción [founding Mercosur]. Today, we're talking about integration."

The national security framework articulated by Canese illuminated an anxiety that repeatedly surfaced throughout the three days: how to manage the asymmetrical relationship with Brazil. Deep ambivalence characterized mention of Brazil, resulting in negative comments like "the favela-ization of Paraguay," contrasted with the ever-present need to stay connected to the Brazilian economy, highlighted via compensation. To resolve the tension, the Hydroelectric Commission's strategy looked to tighter affiliation with other South American countries as a counterbalance to Brazil. The strength of the Hydroelectric Commission's approach was also perhaps a liability: because they focused so tightly within Paraguay, dynamics within the region were overlooked or oversimplified. For example, electricity sales to Chile and Uruguay depended on transmission through Argentina. But this left

Paraguayan energy sales at the mercy of terse international dynamics between Argentina and Chile and between Argentina and Uruguay, to say nothing of volatile politics within Argentina.

So determined was the Hydroelectric Commission to disseminate information about the additional Itaipú revenue and energy integration that, in addition to distributing CD-ROMs and bound reports of all the presentations (for free), they created a government website (now taken down) where hundreds of pages of information on Paraguay's legal codes, development needs, energy market analyses for the continent, and all the content from all the seminars and public debates might be freely downloaded. While Internet access in Paraguay is limited, the combined effort of printed publications, online information, and meetings held in municipalities throughout the country was intended to increase the sense of ownership and transparency across the citizenry.

As Itaipú was an allegory for good government and prosperity, it seems that what was wrong with the past was that Paraguayan elites allied with foreign business interests and, instead of developing the country for the people, developed it according to the profit motive for the benefit of capital. The remedy was a state in defense of the citizenry, investing in and (partially) directing the economy with an emphasis on construction projects and increased energy exports as the main drivers of growth. In spite of the cautions against Dutch Disease, however, this economic model lent itself to some of the same vulnerabilities by leaving most internal growth to unskilled sectors where demand for services would be limited by incomes themselves dependent on unskilled labor opportunities. That is, the wages of construction workers have capped growth potential. And so, the risk was that without a concrete plan to invest in human capital, an export-oriented energy industry (with little room for job growth) could exacerbate the division, creating a loop where energy rents would be invested primarily in lower-skilled sectors rather than in higher-skilled domestic industry. Any increased buying power from economic growth would be then satisfied with imported goods. The problem here would be that the new model, in keeping with critiques of neoextractivism, in effect maintained the frailty of the agro-export economic model, replacing soy with electricity, without ensuring against the immiserating, deskilling effects of growth led by primary resource export.

For the Hydroelectric Commission, the ideal type of government to succeed the ousted Colorado machine was one from the grassroots up, a participatory democracy that not only represented but emanated from the citizenry. This accompanied an international vision of popular sectors united across national boundaries, demanding government action that met the expectations and interests of the citizenry and not just a ruling elite. By participating in political decisions, Paraguay's citizenry would mature beyond the patron-client system of the past. Employment would come as a corollary of state-led expenditures and education out of awareness-raising experiences with the state. Though the new socialism of Latin America in the twenty-first century proclaimed energy sold at market-based prices as fair, it was still the state that would usher in the proper distribution of resources and the right disposition toward the proper relationship of nation to state. The participatory lawmaking of the Hydroelectric Commission was how the state would be occupied by the people to serve a popular agenda of development. But since the leftist vision of a post–Colorado Party nation-state had to contend not just with the liberal technocratic one discussed next, but also with the lively remnants of a political economic apparatus that perdured for more than six decades, this process proved fraught with perhaps irreconcilable difficulty.

Itaipú Executive Directorate: Representative Democracy, Associated Development, and Deeper Integration with Brazil

Not dissuaded by their sidelining in the final stages of the Joint Declaration, the Itaipú Executive Directorate continued to pursue their development agenda. Whereas the Hydroelectric Commission focused on energy rent investments and infrastructure improvements, the technocratic team highlighted the transformative role of participating in an efficient, market-driven industrialization as the way to bring economic and political transformation. During the executive directorship of Carlos Mateo Balmelli (2008–10), a consistent pitch was made to Brazilian (and other foreign) audiences inviting direct industrial investment coupled with the metamorphosis of ANDE from mere producer of energy sold in Brazil into an international electricity firm, with the financial and structural wherewithal to compete on the Brazilian market. The resource curse effects of concern were cast as underdevelopment and weak institutionality. And so, while the energy experts assembled by the Hydroelectric Commission spoke of the need to cure Dutch Disease through government-led financial investments administered via law, the Itaipú technocrats sought to triangulate the danger of Brazil's monopsony power by being *in* the Brazilian market.

Electrification was only part of a multipronged strategy planned by the Itaipú technocrats. The Executive Directorate held integration with Brazil as the most viable linkage for economic development, counting on the increased expansion of Brazil on its ascent as a "global player." They prepared their strategy by reading Brazilian history both to learn from Brazil's model but also to be able to form arguments that would convince a Brazilian audience. Balmelli once referred to an edited volume on the baron of Rio Branco's international diplomacy, written by Brazilian and American scholars, as "an important book for you to read," because it mapped out how the chief diplomat modernized Brazil in the early twentieth century, setting in place trends that continued to the present. At another time, Guillermo Velázquez gave me a digital Portuguese-language copy of Mario Gibson Barboza's memoirs. In *Na Diplomacia, o Traço Todo da Vida* former Brazilian foreign minister Barboza recounted his version of the two agreements, the 1966 act and the 1973 treaty, that established Itaipú.[11] The Executive Directorate team even read reviews of the US-Canada Columbia River Treaty (1964) in order to understand the implications for cross border hydroelectric cooperation. And, yes, they looked to Arendt, Brazilian legal expert Laercio Betiol, American economists Werner Baer and Melissa Birch, and American historian Steven Topik, among others. The preparation paid off, helping the team craft a message legible enough to find traction with Brazil's political and economic leaders.

The Federation of Industries of the State of São Paulo (FIESP) invited Balmelli and Brazilian executive director Jorge Samek to a meeting in May 2009 in order to present their case in a debate format similar to Parlasur. More than 130 trade associations are represented by FIESP, making it the barometer of Brazilian industry and arguably the most powerful business organization in Brazil. "We are in a process of integration, commercially, and above all, energy integration," Balmelli said to the FIESP audience. "There is one country in the region that is a *global player*. . . . Will it be a *global player* representing South America with South America or without it? . . . Does Brazil want associated development or dependent development?"

Balmelli pressed his listeners toward an economic integration built on "associated development" and deeper Brazilian commitments with its neighbors. The presentation ended with vigorous questioning. However, Balmelli's appeal to financial interest found favor. Having a poor neighbor with little buying power, he said, was bad for business. Balmelli advanced an analytical, not merely rhetorical, framework of associated versus dependent development and the one path to modernization that could include the people in Paraguay, electrification. If ranching and agribusiness were exclusive, then electrification had the opposite potential, one to bring more people into higher-skilled, higher-waged industry and thus reach those who had previously been left out of economic growth. Balmelli advocated for a manufacturing industrialization because of its higher need for manual labor.

The terms "associated development" and "dependent development" (*desarrollo asociado, desarrollo dependiente*) hearken to Dependency Theory analyses of Latin American economists like Raúl Prebisch in the 1950s and, especially, Fernando Henrique Cardoso in the 1960s and 1970s.[12] Briefly, Dependency Theory attends to the core-periphery relationships within the modern capitalist system, arguing that inequalities are currently structured into the system by how the core depends on the resources of the periphery. Cardoso, in fact, coauthored an article entitled "The Contradictions of Associated Development" where he argued for a nuanced understanding of how state-involved companies positively interacted with foreign companies and the local private sector to yield economic growth, using Brazil of the 1960s and early 1970s as a case study.[13] In posing the question of associated versus dependent development to FIESP, Balmelli verbally invoked the polices of Lula da Silva's presidential predecessor Fernando Henrique Cardoso and the economic growth experienced when Cardoso's Plan Real stabilized the economy. Associated development sees a jumping of scales as the means to sustainable economic growth, and it prominently features state-owned corporations as a necessary component. This was no neoliberal orthodoxy. Rather, the Itaipú team advocated for public-private partnerships linking Paraguay's state-owned energy company to private Brazilian business in order to foster industrial growth.

Just as the economic priorities differed between the two Paraguayan energy teams, so too did their communications strategies for how to get those plans implemented, which indicates how they understood state power to work. For more than three years of Lugo's presidency, the Paraguayan government lacked ambassadorial representation in Brazil, a situation not remedied until March 2012. It seems that the priorities of the Ministry of Foreign Relations lay elsewhere as the Hydroelectric Commission toured Paraguay and solicited input from outside experts. Balmelli and his close team of advisers, on the other hand, consistently lobbied ("*hacer* lobby") powerbrokers in Brazil. After having been removed from the final negotiations in July, they embarked on almost weekly trips to Brazil to meet with the leaders of a broad range of political parties in congress and to speak to business and industry moguls about how to invest in Paraguay once its energy infrastructure improved. What had only been mentioned indirectly before the Joint Declaration became the core platform of Itaipú-based lobbying in Brazil: to attract Brazilian industry to Paraguay through three incentives, cheap energy from Itaipú (and possibly other dams), cheap labor costs, and a low tax burden. This message was presented by Balmelli and his team to audiences of business elites throughout Brazil in the second half of 2009. Balmelli was invited by FIESP back to São Paulo in October 2009, during which he was able to privately meet with the man he had cited months before.

"Integration is the destiny of our peoples," said Fernando Henrique Cardoso after his conversation with the Paraguayan executive director.[14] The former president of Brazil, in addition to his scholarly expertise on Latin America economics, was a senior member of the second-largest opposition party in Brazil, the Brazilian Social Democracy Party (PSDB). Their favorable votes would be key in passing the Joint Declaration in congress. In November, Balmelli visited a prominent senator from Lula da Silva's own Worker's Party (PT). News of the conversation spread through Brazil, where PT senator Delcídio Amaral spoke warmly of the Joint Declaration's potential to bring development to Paraguay and praised Balmelli as "one of the best cadres of Lugo's government."[15] After his conversation with Lula's ally, Balmelli met with other leaders of Lula's opposition, Sérgio Guerra of Cardoso's Brazilian Social Democracy Party and Romero Jucá of the centrist Brazilian Democratic Movement Party.

The Brazilian press described the Executive Directorate's actions in football terms, "Paraguay takes the offensive to sell energy," and referred to Balmelli himself as "the principal negotiator on the Paraguayan side."[16] In spite of Lula da Silva's support of the Joint Declaration, strong reticence to the deal lingered within Brazil, such that Balmelli had to say, "We do not come to commit terrorism within the local energy market."[17] The Itaipú technocrats proposed to immediately begin building ANDE's capacity to sell at the market price rather than wait until 2023 to sell to third parties. Behind the scenes conversations between the Executive Directorate and the leadership of ANDE centered on the financial mechanics of increasing the Paraguayan utility's purchasing power gradually but steadily, since ANDE would need to buy electricity from Itaipú to then sell it on the Brazilian market.

The directorate also traveled to the other main industrial center of Brazil, the south. Seated between Itaipú Brazilian executive director Jorge Samek and the Brazilian ambassador Eduardo dos Santos, Balmelli made his case to the American Chamber of Commerce in Curitiba (AMCHAM) on November 16, 2009. Again, the diplomatic imbalance was striking. At AMCHAM, Balmelli argued that, in order to bring "real economic" growth based on production, Paraguay needed the direct investment of Brazilian industry in Paraguay.[18] Reiterating the speech he had made almost a year before at Parlasur in Montevideo, he affirmed that Brazil was on the ascent and that Paraguay should link itself even more closely to its neighbor's emergent rise. The plan was to use Itaipú as a point of departure, sheltered from previous patterns of Paraguayan political and economic inefficiency because of the dam's extrajuridicality.

The Itaipú Directorate expected more was to be done and won in Paraguay's negotiations with Brazil—that the powerbrokers were not just the president, but the legislature that would have to sign an unpopular agreement. Their actions signaled that they thought this might be more palatable if the business community saw an opportunity for investment where energy and labor would be cheaper than in Brazil. Note that the Executive Directorate, in fact, prioritized political parties of the opposition. Balmelli's team of technocrats openly advocated for a form of government-led development that focused on attracting industry and investment, expanding energy consumption, and removing partisan or personalist criteria for projects.

To this end, the Itaipú Technology Park (ITP) was revamped on the Paraguayan side. In 2003, the ITP was established on both the Brazilian and the Paraguayan sides of the dam complex. Per the 1973 treaty, equal monies were disbursed to both halves, but by my first 2008 visit, the state of the park on the two sides was incomparable. In Brazil, we were shown advanced biotechnology laboratories where vaccines were being manufactured and a nascent pharmaceutical industry (a public-private partnership) was working to develop new drugs. These industries were so important within the Brazilian government that they were referenced as a matter of national security in a Brazilian National Defense Strategy working paper signed by Lula on December 18, 2008. Jorge Samek himself led our small group to sterilized advanced computer programming rooms and to well-equipped high-tech classrooms for teaching computer engineering at the new university being built in the ITP, intended to be the MIT of Latin America. On the Paraguayan side, empty doorless buildings poked out from wild jungle overgrowth, red mud splattered on the walls with no indication of how the millions of dollars distributed to the Paraguayan ITP had been spent.

One year later, the Paraguayan ITP was being planned as a technology research facility to employ US-educated Paraguayan STEM PhDs whose scholarships were to be funded by Itaipú and the Fulbright commission. And the site was being shopped to potential Brazilian investors as the future location of the maquila industry in Paraguay, using the proximity of the border department of Alto Paraná as well as the ease of access via waterways to distribute goods to the Brazilian market. Appended to this were infrastructural projects throughout Paraguay. Not only was the five-hundred-kilovolt transmission line planned as construction on the Paraguayan Itaipú substation continued, but the engineers on the team proposed to build dams on Paraguay's internal rivers to commercialize this energy in Brazil without the controversy of Itaipú and the Joint Declaration (including a plan to dam the Monday River, which possesses the impressive Monday Falls).

The ITP exemplified associated development as written about by Cardoso, but with a modern twist. South America had already experienced the economic growth promises and disappointments of neoliberal orthodoxy, which pledged to wipe away poverty by the privatization of government companies and services. While Itaipú technocrats championed market-led growth, their trust in just any market was qualified: GDP and economic sector growth could be exclusive when they accompanied the creation of an export enclave. Development required a paternalistically involved state in order to ensure that economic growth was inclusive. The twenty-first-century touch was the prominent role of regional integration, not merely public-private partnerships, as part of the strategy—in the maquila sector, in energy sales, in foreign direct investment.

Participatory democracy was not the only model for changed governance advocated for by members of the post-Colorado government; the Executive Directorate promoted representative democracy. A new breed of politicians across several parties called for a rationalization of politics according to modern, liberal democratic lines in the tradition of Brazil's technocratic expertise. In Itaipú, the circle of advisers around Balmelli hoped to see technical qualifications and rational expediency as the basis for how the dam should be administered, where skilled and principled leadership was the lynchpin for a state that did right by its citizenry.

Development of this kind placed decision-making power in the hands of technocratic elites who had specialized training in business and finance and who would directly persuade foreign investors. Only the select few who had access to Balmelli's schedule could make suggestions as to what the needs on the ground might be and how best to address them. Underlying the image of development here was the expectation that modernity would come through the experience of working in efficient, market-driven industries and through the education required for employment in technologically advanced sectors. The clientelistic past—a feature of all political parties, not just the Colorado Party—was the enemy to progress. Because patrimonial customs pervaded most government institutions and even Paraguayan business interests, direct foreign investment in an area headed by a team of market-oriented experts would be the only way to circumvent rent-seeking behavior. Aside from questions about whether a maquila-based industrialization oriented toward the Brazilian market would successfully bring development, another debility of this top-down decision-making process is that it replicated the practice of the previous system, leaving those most affected by decisions unaware of how they came to be made. While the goal might be different, because the method looked similar, it lent itself easily to accusations of the same excesses of previous governments and reinforced the expectation of patronage that the new Lugo appointees in Itaipú saw as the obstacle.

The entire situation laid bare very different notions within the Paraguayan government of how to treat with Brazil and of how power is distributed within the Brazilian government and within the international state system. Lugo, the Hydroelectric Commission, and much of the Paraguayan government (not just his progressive base) acted as if the main gatekeeper was the president of Brazil; once the document was signed, the Joint Declaration had only to be ratified pro forma by Congress. The Paraguayan experience of government holds that formal, public signings happen only after decisions have been made in private meetings and after assurance of necessary votes has been gotten. And in Paraguay—especially, but not only, among popular sectors—the threat of international arbitration was conceived as a realistic escalation measure to force Brazil's hand. Perhaps because the progressive campaign had focused so much on making Itaipú a national cause, the international character of the dilemma seemed less immediate.

The need to continually petition political and economic power holders in Brazil for passage of the Joint Declaration was almost exclusively the priority of the Itaipú technocratic team, and many people I spoke to in Paraguay did not know the country lacked an ambassador to Brazil at the time (although, obviously, the Hydroelectric Commission was aware). By acting as the de facto advocate for Paraguay and the adoption of the Joint Declaration during weekly trips to Brazil, Balmelli evinced a very different understanding of political pressures within Brazil. He and his team of technocratic advisers considered individual political parties and business sectors to be stakeholders whose consent would be necessary and so stepped-up lobbying to these audiences. Here there was no notion that any kind of threat would work in Paraguay's favor, only self-interest and the promise of economic benefit through industry.

Balmelli tenaciously lobbied Brazilian politicians and industrialists. In January 2010, as he visited with the president of the Brazilian Socialist Party in Recife, he received an unexpected phone call from one of Fernando Lugo's underlings. He had been fired.

Conclusion: Both Roads Not Taken

In Balmelli's stead, Lugo appointed leftist Gustavo Codas, who immediately declared that his administration would work much more closely with municipalities and governorships in order bring development. Over the course of Codas's tenure as interim executive director (2010–12), revenue transfers to municipal and departmental governments increased by more than 200 percent under the Royalty Law that was only a few years old at the time (see chapter 2). No formal reason was given for Balmelli's dismissal. Balmelli attributed it to an anticorruption case he had initiated against one of Lugo's other appointed Itaipú directors who had publically admitted to falsifying paperwork to direct Itaipú dollars to the Colorado Party election. (In spite of the public confession and physical evidence, the official remained in his position until being transferred to head another Itaipú division.) Others claimed it had to do with Balmelli's "arrogant leadership," as was said repeatedly to me by leftists and in the press, or Lugo's desire to prepare for the midterm elections by putting a closer collaborator in charge of Itaipú. I suspect it had as much to do with eliminating a rival who, through the international legitimacy he was attaining in negotiations, was consolidating an alternative source of domestic power. The Liberal Party cried out in outrage that one of their own had been replaced by a leftist sympathizer. By nightfall, Lugo announced that a member of his vice president's Liberal Party faction was the new person in charge of Yacyretá Dam, and a new Liberal Party appointee, later accused of millions of dollars of embezzlement, was named head of ANDE. Umbrage turned to celebration.

Now holding the reins at Itaipú, the Hydroelectric Commission began to adopt some of the former Itaipú Directorate rhetoric. Codas made a pitch for industrializing via electrification to the American-Paraguayan Chamber of Commerce in April 2010. He even spoke of attracting investment and foreign industries. And he set the groundwork for a potential multimillion-dollar deal with Rio Tinto Alcan, a Canadian aluminum company, which contracted his brother Ricardo Codas as a formal representative in Paraguay and would use more than a turbine's worth of Itaipú electricity at very low rates to manufacture aluminum from bauxite. The deal fell through a few years later, after having attracted the attention of environmental groups because of the pollution impacts of aluminum processing.[19] But progress on the Joint Declaration slowed for nearly two years. Lugo's closest allies in the Ministry of Foreign Relations exerted less effort to convince the Brazilian congress to approve the changes. Instead, Lugo and leftist social movements turned their attention to preparing for the 2010 midterm elections. After Balmelli's dismissal, Paraguayan efforts to lobby Brazilian elected officials diminished into a few exceptional moments many months later, a trip by the minister of foreign relations, and a trip by a few senators. But the Joint Declaration remained unapproved and the pledged financial benefits still only a paper promise. The midterm elections saw massive gains for the Colorado Party. As time passed, grumbling began in Paraguay that Lula had lied or that Brazil at large had no intention of fulfilling the declaration. And when it was announced that it was unclear whether Itaipú or Brazil would finance the construction of the five-hundred-kilovolt line (it was eventually financed by Mercosur) and that the one-year anniversary of signing the agreement would pass without even a vote in Brazil's congress, the tenor of the complaints in Paraguay grew bitter.

The year 2010 came and went with no advance on the Joint Declaration. No ambassador to Brazil was named, and Lula stepped down from office to be replaced by Dilma Rousseff, his chosen successor who nevertheless lacked his popularity. The Brazilian congress repeatedly postponed discussion of the amendments. Rather than beginning to receive the increase in energy payments by the end of 2009, Paraguay instead saw 2011 begin without a planned vote and no additional income. It was not until April 6, 2011, that the Brazilian lower chamber and May 11, 2011, that the senate approved the amendments that ratified the Joint Declaration. The first additional compensation payment finally arrived in October 2011. But even this was not to the liking of the Hydroelectric Commission; the Paraguayan congress ignored the proposed Development Fund and instead transferred 50 percent of the new proceeds directly to municipal and departmental governments to spend, presumably, on recurring costs. Because of the rift with the Liberal Party, Lugo's governing coalition did not possess enough votes in congress to neutralize the spending plan.

Congress passed a law establishing the National Fund for Public Investment and Development (FONACIDE) to administer all the compensation

TABLE 5.3. From the National Fund for Public Investment and Development

a) 28% to national treasury for programs and infrastructure projects;
b) 30% to the Excellence in Education and Research Fund;
c. 25% to departmental and municipal governments;
d) 7% to the Financial Agency for Development for capitalization;
e) 10% to the National Health Fund.

income one year later. (Royalties were administered by Hacienda under the Royalty Law.) Using some of the idioms of the Hydroelectric Commission bill, FONACIDE nevertheless focused (almost) exclusively on rent redistribution to short-term projects or direct cash transfers (see table 5.3). One quarter of the funds were transferred directly to municipal and departmental governments (in addition to the royalties already distributed to eastern departments that bordered the Paraná River). Education funds were directed by FONACIDE to school equipment, infrastructure, meals, and research competitions, and health funds to medicine and equipment. It was unclear from the provisions of the law whether any infrastructural projects were intended to increase electricity access, and the programs to be funded by the national treasury were left similarly unspecified.

In other words, FONACIDE provided revenue to be redistributed to recurring expenses in lieu of a tax base or to be apportioned according to clientelistic networks. If the Hydroelectric Commission team wanted to counteract Dutch Disease through long-term financial investments that would facilitate electricity consumption and if the Itaipú technocrats sought to pair development within Paraguay with the transformation of ANDE into an international electricity firm, the Paraguayan government instead returned to Itaipú as a source of rent. Nothing in the text of the law establishing FONACIDE acknowledged that 2023 would entail a renegotiation of energy pricing and even the concepts of royalties and compensation themselves. And, as might be expected, the longer but more permanent endpoint of the dam's future obsolescence was unmentioned. Despite warnings from financial experts that municipalities receiving 40 percent of their income from binational cash transfers caused a deformity, the post-Lugo policy put that trend on steroids. In 2016, Royalties and Compensation provided municipalities with 67 percent of what they received from the central government; FONACIDE provided another 26 percent, meaning that 93 percent of transfers from the central govenment to municipalties came from energy rent. That same year, departmental governments got 10 percent of their budgeted transfers from FONACIDE and another 20 percent from Royalties and Compensation for a total of 30 percent.[20]

Whereas the Joint Declaration had been described in 2009 as the end of the War of the Triple Alliance, the slowdown in Brazil's congress was explained as yet another instance of patterns seen in that war. This was Brazil, again, casting its imperialist weight around, using deceit to get whatever it wanted out of Paraguay, and never intending to keep its promises. Perhaps, however, the stalemate arose not because of Brazilian deceit but, in part, out of the choices arising from Paraguayan government assumptions about how political decisions are made and implemented in Brazil. Perhaps the international strategy of Lugo's government—to assume that Lula's signature on the document assured its passage in Congress and that no ambassador or regular advocacy campaign to Brazil's political and economic elites would be needed—was partially responsible. Moreover, when Lugo removed Balmelli from office, he not only eliminated a rival to the Hydroelectric Commission and placed Itaipú in the hands of someone more closely affiliated with the Paraguayan left. Lugo also discarded the chief interlocutor trusted by Brazilian politicians and industrialists. And he lost the one Liberal in his government who could counteract the machinations of the Liberal Party vice president and who could garner support from more traditional sectors within the Paraguayan government. Nevertheless, the negative experiences from the aftermath of the Joint Declaration, rather than calling into question the foreign policy strategy, merely psychologically reinforced the master narrative that this was a continuation of the War of the Triple Alliance, Brazil looking for a way to take advantage of Paraguay.

The Hydroelectric Commission and the Itaipú Executive Directorate offered two potential models for development based on exploiting hydroelectricity; the fact that both involved state- and market-led growth is not just a Latin American peculiarity. Rather, the central role of natural resources of the national territory calls for state actors and institutions to steward the patrimony of the nation, and hydrodollar dynamics entail political economic arrangements that foreground government interventions. The Hydroelectric Commission attended to the aspects of the Joint Declaration that resulted in immediate financial transfers to Paraguay that would be administered by the executive branch. Just as hydroelectric sovereignty was a new technopolitical subjecthood, so, too was Itaipú as a so-called border plant an attempted reterritorialization of power. On the other hand, the Executive Directorate looked to move away from redistributed income and to build capacity—the financial capacity of ANDE, the manufacturing capacity of

Paraguay, and the educational capacity of the citizenry. Neither of the proposals was implemented, but not because the models were evaluated and discarded. Instead, they were cast aside because of fierce partisan dynamics within Paraguay and because rent was an easy and established course to follow.

Fernando Lugo himself was impeached on June 22, 2012, in a surprise trial on trumped-up charges of fomenting rebellion following a deadly gunfight between police and peasants in eastern Paraguay. The backroom deal involved Colorados, Liberals, and other centrists who waited well until the Itaipú matter was settled to enact their plan. All of Lugo's closest associates were summarily dismissed as new president Federico Franco installed his favorites in ministerial positions where they could directly access government coffers; his administration passed the FONACIDE law. The governments of the region universally denounced the impeachment as a coup and suspended diplomatic relations with Paraguay, which were reinstated upon the election of a new president from the Colorado Party the following year. By this point, the Joint Declaration's gains had trickled down to direct cash transfers via compensation. Most of the pledged construction projects stalled. Mention of ANDE electricity sales on the Brazilian market vanished entirely. Itaipú broke world production records in 2012, and the Paraguayan government received record amounts of energy rent.

6

Ecoterritorial Turns

Introduction

DECEMBER 2008

Paraguayan poet Areté paused to translate her own work in the middle of the poetry reading. Having returned from a fellowship in Japan, she was at the top floor of a downtown Asunción gallery to launch a book of haiku composed in Guaraní. But because she knew her urban (and international) audience might not command the indigenous language, she recited, "El agua nos divide, la tierra también. El cielo nos une."

The water divides us. The earth, as well. The heavens unite us.

Areté stands in a long tradition of Paraguayan authors who wield the spoken and written word as a way to express hope and defiance. The artists of the landlocked country are its most important interpreters of politics, history, and economy. Amid fears that arise over globally desirable natural resources, the imagined threats and risks that get attached to water are, in part, remedied via political solutions. As Areté presciently called out, the story of Itaipú and renewable energy in South America thus far has been one of how water and land divide. But in this chapter, we will explore how water creates community. More precisely, we will see citizenship as constructed in relation to transboundary water, distinct from patterns of governance based on fixed territorial boundaries, as we look to connections between legal priorities, climate change, and the durability of capitalism. In liberal democracies, communal hopes and fears get fought over in law. Here I take law as a site of values enforcement through two documents, the signed Joint Declaration (2009) and a proposed region-wide South American Energy Integration Treaty, which rescripted sovereignty and state power under new hydraulic pressures.

WATER INSECURITY

Brazil and Paraguay are caught in a double bind as they share the dam. The thickened binational integration forced by the dam regularly triggers sovereignty concerns. Yet integration simultaneously serves as a form of security. In spite of all the engineering interventions possible, Itaipú Dam's ability to produce current depends on that other current, the flow of the Paraná. For both countries, the cyclical reliability of electricity and energy rent translate energy security as national security by bringing money to elites, by bringing energy to capital, by putting infrastructure on the unruly hinterland, and by determining how rivals adjudicate water. Integration also functions as national security vis-à-vis natural resources in light of global anxieties around water. Given that variation in precipitation jeopardizes the entire logic of a hydroelectric system, anthropogenic climate change threats occupied a curious part of the larger discourse of hydroelectric and hydraulic anxieties in Paraguay. They were surprisingly muted. Engineers spoke of rainfall, but not of the possible changes to the predictable movement of water that might come with extreme weather. At no time did I hear Itaipú energy managers or the Hydroelectric Commission mention what they might do should rainfall patterns change, impeding the ability of the dam to produce at its historical rate. Instead, the concern was that climate change and scarce resources elsewhere would threaten the security of water-abundant environments.

There is something about water that distinguishes it from other attributes of nature and the national territory. Yes, other forms of energy extraction have had landscape-altering effects, many of them terribly destructive. Yes, national-territory-as-patrimony is a common framing in Latin America, with the attendant notions of destiny and inheritance, where nature is a gift or a trust that is to be shepherded by the state in fulfilling the destiny and development of the nation. Nevertheless, water stands apart. Perhaps this is because we have not yet been able to contain the destructive power of water: the Indian Ocean tsunami of 2004, hurricanes Katrina and Maria. Our imagery of photogenic natural disaster and devastation wrought by anthropogenic climate change is frequently charismatic water imagery: the dried Aral Sea littered with carcasses of ships, retreating glaciers on Kilimanjaro. Perhaps, though, the root of the matter is that we must drink water daily for our survival—not just use it for agriculture, industry, or hygiene. The visceral, personal connection drives our relationship to it.

This may add to the anxieties experienced in the larger River Plate basin in which the Paraná River and Itaipú are located. For as much as management of the binational hydroelectric resources of the Paraná have unfolded into an unprecedented level of regional cooperation in the Southern Cone, much more is at stake. Even as the EU's legal genealogy may be traced to coal and steel agreements between France and West Germany in the aftermath of World War II, Mercosur has its origins in the administration of international water resources. Whereas international organizations (IOs) were established in the North Atlantic to facilitate international diplomacy or global economics (e.g., the UN, IMF, World Bank), Itaipú was an IO designed to manage nature. An "IWO" (international water organization) or an "INRO" (international natural resource organization) such as Itaipú does more than just set environmental policy or recommend action, as the Kyoto Protocol might. The materiality of the dam, of the energy, and of the water is at the very center of what is managed. Because of climate change, the converse may also be true: through an international national resource organization, politics and economics may be managed by water. If "citizenship" may be a way to describe the responsibilities and connections implied in the hyphen that links "nation" to "state" in "nation-state," the obligations of energy or ecological citizenships demonstrate that differing relationships to nature result in different kinds of political claims, testament to the rising importance of environmental elements in governance.[1] Not only has Mercosur become the common market of choice in South America; Itaipú and the broader administration of the River Plate basin are an antecedent for managing an even more important water resource: the Guaraní Aquifer, a massive source of ground water that drains much of the best farmland in Argentina, southern Brazil, Paraguay, and Uruguay. And if sovereignty can be descried in the flowing of water to refill a reservoir, then it is all the more complicated by the dynamics of a transboundary aquifer.

I am not the first to make this connection. Possession of desirable resources elicited consternation in Paraguay beyond regional balances of power or Brazilian encroachment. Early in my fieldwork, when I explained my research interest in the region's water resources, the eldest son of a high-ranking political family in Paraguay took me aside. He sternly cautioned me,

"Can I give you some advice? Never, ever mention the aquifer again. We will think that you are going to invade to steal it" (personal communication, December 2008). In Paraguay, water insecurity was not imagined as drought. Instead, water insecurity was configured as a militarized intervention to physically control or even remove the resource, an anxiety that harkened as much to the recent Bolivian water wars (1999–2000) over privatization of Cochabamba's water system, but also the fears arising from the trauma of the War of the Triple Alliance.

The wall of the Asunción office of the Paraguayan executive director of Itaipú featured, in addition to a large aerial photograph of the dam, a large map of the Guaraní Aquifer with discharge and recharge zones (and national boundaries) clearly labeled. And at an extraordinary meeting of the Mercosur parliament in December 2008, summoned for an update on a diplomatic crisis between Paraguay and Brazil over Itaipú, one senator from Uruguay closed his comments (in support of the Paraguayan position) by saying, "We will have the Guaraní Aquifer; we have Bolivian natural gas; there are a dozen issues that are the assets possessed by this South America. I believe that we have the unavoidable responsibility to find an adequate solution for these matters."[2] The resolution of binational conflicts within Itaipú was understood to be a dress rehearsal for the aquifer "possessed by this South America" and, I assert, has direct implications on that other continental asset, what is arguably the world's most important body of freshwater: the Amazon. As a major carbon sink and clean air provider, the Amazon has direct worldwide impacts and pits the local, national, regional, and global against one another. Because technoscience can act as seemingly neutral political leverage in law, threats of environmental crisis possess the ability to destabilize universal liberal values.[3] Ecological disaster and anthropogenic climate change effects are not limited by national-state boundaries, rescaling ecodangers as a universalizable threat and consequently giving added weight to ecoresponses.

A larger ecoterritorial turn in twenty-first-century South American politics, in which the national territory in the national imaginary is strongly marked by the ecological content of that territory, can also be seen in jurisgenerative encounters around Itaipú Dam. Law is made from the field of fraught contestation comprising political struggles between the Right and the Left, emergent energy integrations versus nationalist agendas, and crises with variant temporalities. That the governance of (and with) institutions that manage ecology should, as we will see, be so productive of political-legal forms helps answer the question posed in the introductory chapter: how to apprehend the place of energy capture in sociocultural formations.

Rather than a unidirectional determinism, energy and society are dialectically linked via a metabolic relationship that transforms them both: dams change landscapes; the properties of water change states. With meanings that are neither obvious nor predestined, the linkages are complex and highly interpreted. Indeed, how else should hydroelectric potential signal memory of war?

International agreements and treaties compress time and space, drawing energy from those elements held together. Amid all the possible points of convergence, of greatest interest in this chapter is how rights get attached to water-as-energy and how rights are generated by water-as-energy. Because of the quality of movement, water's ecoterritorial attributes exceed the boundaries of the national state, implying a larger region as the basis of an ecocitizenship. We have looked for the contest over power in turbines, currency denominations, job titles, the location of a conference table; here we find how ethics are expressed through renewable energy as law is produced and circulated through Itaipú. The direct lineage, for example, that can be traced between the founding of Mercosur and the water/energy agreements around Itaipú Binational Dam explains the greater significance of the Lugo-Lula Joint Declaration (2009), in spite of how it was implemented. The Joint Declaration was merely one of several recent agreements treating water in the Southern Cone that initiated new developments in rights, sovereignty, and integration. And in the flow of law, Itaipú is a converter substation that changes the frequency of international relationships. I argue that the bilateral Joint Declaration (2009) and a multilateral push for a South American Energy Integration Treaty, which was signaled by the Joint Declaration, are two exemplars of how water leads to politics in the context of new anxieties around the resilience of nature, liberal democracy, and economic liberalism.

Three perceived threats have imperiled the global capitalist political economic system and its incarnation in South America, which both the Joint Declaration and the Energy Integration Treaty seek to ameliorate through management of energy as a means to defend political-economic systems, exemplifying what is meant by "ecoterritorial." The first is a temporal problem of natural resource scarcity and of peak oil, exacerbated by anthropogenic climate change and by international climate accords like the 2015 Paris Climate Agreement (COP 21) requiring the movement away from fossil fuels. The second is a conceptual problem about how to envision proper relationship to nature. Over the past five hundred years, Western governance has been characterized by increasing attention to rights of individuals as understood in Enlightenment thought: life, liberty, property, civil and human rights, citizenship. But in the early twenty-first century, nature and natural formations are increasingly the subject, not just object, of rights.

The last two decades have seen a political resurgence of indigenous cosmologies and values regarding nature that implicitly and explicitly challenge extractivist approaches to resources attendant with capitalist market expansion by recognizing the personhood of nonhuman and even inorganic things.[4] Perhaps the most significant development in Latin America was the recognition in the 2008 Ecuadorian constitution (articles 14, 71–74, and 414) of the inalienable rights of nature to "exist, persist, maintain and regenerate its vital cycles, structure, functions" (article 71). Bolivia has also ratified a Law of Mother Earth (Law 071, 2010), which rests on the principles of "harmony, collective good, the guarantee of regeneration, respect and defense of the rights of Mother Earth, no commercialization, . . . cross-culturalism." The government of New Zealand recently recognized the Whanganui River a living entity and was quickly followed by the government of India, which declared that the sacred Ganges and Yamuna Rivers had the legal status of a person, with all the appropriate protections under the law.

A third anxiety is similar to the one voiced by the Paraguayan dinner guest when I mentioned my general interest in water resources and the Guaraní Aquifer: a jurisdictional problem of sovereignty. Government officials (and ordinary citizens) of the Southern Cone mentioned the Guaraní Aquifer as the logical endpoint of whatever happened with the Itaipú controversy during the Lugo government. This was why the Uruguayan senator openly referred to the body of fresh groundwater at Parlasur in 2008 and on the wall of the Itaipú executive director's office in Asunción hung a detailed map of Guaraní Aquifer. The fact that neither the Guaraní Aquifer nor the Amazon River sit within the boundaries of any particular nation-state creates a conundrum of whose water it is and how to rein in upstream activities that detrimentally affect downstream communities. Moreover, in a time of scarcity and predicted water wars, the worry is that countries in the Global North will decide to seize the water resources. And in any case, there is a perception that the old structures that adjudicated decision making over transboundary water resources might not be apt to climate and political conditions of the present.

Arbitration of transboundary water rights and obligations in South America has shifted over the past sixty years from conflict to cooperation. As the new legal framework built around Brazil and Paraguay unfurled, Argentina had a prominent seat because rapprochement between Brazil and Paraguay represented a Paraguayan pendular swing away from tight association with Argentina. More importantly, Argentina lies downriver of Brazil and Paraguay. Its *aguas abajo*, downstream waters, are affected by activities *aguas arriba*, upstream. In the middle of the twentieth century, dam building dominated international concerns in the Southern Cone, but in the present, these have been eclipsed by the Guaraní Aquifer's "recharge" and "discharge" zones. Those areas where water may enter (recharge) the aquifer lie almost exclusively within the national territories of Brazil and Paraguay, leaving Argentina and Uruguay with sites where water might be drawn (discharge), thus placing the agricultural productivity of the latter countries (downstream) at the mercy of upstream countries' policies and of changes in rainfall patterns.

But the metaphor also works for law. Mercosur lies downriver from Itaipú in terms of treaty genealogies and legal-historical context.[5] During the initial Itaipú planning, a 1969 treaty brought together all countries sharing the River Plate basin (Argentina, Bolivia, Brazil, Paraguay, Uruguay) and asserted the rights of the signatories to make bilateral agreements over international waters without the approval of downriver countries but also affirmed "good neighborliness," that such agreements should not cause harm to downriver countries. The arrangement departed from previous water expectations that had demanded consultation with downriver countries, not merely consideration of them. It also hinted at something that would become more concrete in the following century: conceptualizing a political territory as circumscribed not by national borders, but by the boundaries of a basin.[6] And so, international water agreements over dynamics between *aguas arriba* and *aguas abajo* countries flowed into the Common Market.

The intellectual-legal architecture built around energy in both the Joint Declaration and the Energy Integration Treaty, even as it points to twenty-first-century contests over water, provides new ways to think sovereignty versus security and upstream versus downstream relationships. Dry land and immobile objects more easily lend themselves as the targets of political-legal fixity, but water—not just its presence, but its movement—is also productive of political scale as fixed hydraulic and now hydroelectric infrastructure as well as specific rules of access to those infrastructures have provided ways to control water.[7] Access to resources indexes identity and, in fact, partially constitutes identity for both individuals and communities. As more complex thinking on the flows of water arises, one potential area of change pertains to how upstream and downstream communities relate and whether there might be novel conceptualizations of the movement of water and thus access to it (outside the upstream/downstream arrangement) that might lead to new community identities. Though Itaipú sets many precedents in

South America and beyond, this may be the most significant, especially if the Joint Declaration is to the Guaraní Aquifer as the South American Energy Integration Treaty is to the Amazon.

Because the data in this chapter center on a genealogy of legal artifacts and core ideas within them, here I focus on textual analysis, rather than ethnography, in order to tease apart the opening framings of both documents with attention specifically to the rights and rationales that serve as justifications. I attend to citizenship rights first in the Joint Declaration, then in the Energy Integration Treaty, how these are connected to the specific regional context, and what they might mean for water governance in the future. The definitions of community belonging and expectations of the state-to-nation relationship advanced in both documents, that is, the law generated by them, enshrine priorities. Thus the rules of access to resources arise from ethical orientations that are nevertheless conditioned by the fluid mechanics of water. Yet even as arrangements of obligation and membership address portending temporal, conceptual, and jurisdictional crises, a hierarchy of values may pit consumption or wealth acquisition as the consummation of sovereignty and security, over and above other concerns such as human and nonhuman life.

The Joint Declaration as Balancing Act: Economic versus Human Rights, Sovereignty, Self-Determination

The Joint Declaration (2009), product of regional politics with intellectual antecedents, introduced several important elements, particularly in Items 1 and 2, that accrue the potential weight of international law (because they were signed by the two presidents). They read:

> [THE PRESIDENTS OF PARAGUAY AND BRAZIL]
>
> 1) demonstrated their satisfaction with and decided support of the system of representative democracy currently in force in the countries of the region, of the unrestricted respect of human rights, and of the sovereignty and self-determination of the peoples.
> 2) coincided in pointing out the importance of solidarity as a guiding component of the process of regional integration already underway as well as their intentions to continue efforts to diminish poverty, inequality, and other forms of social injustice.

Intended to usher in a new phase of bilateral relations, the declaration begins by recognizing "representative democracy" and "human rights," as well as the "sovereignty and self-determination of the people," as the basis of a process of integration that will diminish poverty, inequality, and all forms of social injustice. These opening statements ground the rest of the declaration in a particularly Latin American context with a legal/philosophical basis following significant jurisprudential trends in postcolonial, postauthoritarian South America. The emphasis on representative rather than participatory democracy hearkens to the difference between the two Paraguayan negotiating teams from chapter 5 and takes the less radical road of the two while levying a criticism against the right-wing authoritarian governments of previous decades. (The subsequent parliamentary impeachments of Fernando Lugo and Lula da Silva's successor Dilma Rousseff, the ascension of unelected presidents and then the election of conservative presidents with explicit ties to the military regimes of both countries,' have illustrated why affirming self-determination and representative democracy is more than a rhetorical gambit.)

The change is significant. Argentine legal scholar Martin Böhmer has argued that the pan–South American experience of the Dirty Wars and brutal right-wing military governments of the 1950s to 1990s at last made an opening for the re-creation of political culture in the continent.[8] An expert in comparative constitutional law, Böhmer maintains that in the founding of the United States the framers of the its constitution feared the power of the many and the one, deciding to put power in the hands of the few, and so the US Supreme Court holds ultimate sway over both the executive and the legislative. In France the framers feared the one and the few but put power in the hands of the many, that is, the parliament. In Latin America, however, the framers feared the many and the few. Thus they sought to create a monarchy dressed as a democracy; in this way Böhmer accounts for the frequently authoritarian presidential system of much of Latin America and the repeated rise of *caudillos* (strongmen) from the early national period to the late twentieth century.

But the military regimes who, in the name of security, detained, tortured, and killed thousands of the citizens they said they were to protect, ruptured the formula "state = law-abiding." Recall that in the 1970s under Operation Condor, joint state terror formed part of the bilateral relationship between Brazil and Paraguay. So severe was the break that a nascent political culture was fashioned where states in South America have had to affirm human rights as the foundation of a democracy.[9] The Joint Declaration's prominent placement of the phrase "the unrestricted respect of human rights"—mention of which would not have happened in the original Itaipú Treaty signed under two military governments—follows in this new direction. Including human rights in the text of the Joint Declaration pointed to years of popular counterhegemonic struggle throughout Latin American civil society for a way to rebut the unchecked necropolitical power of the state to kill citizens with impunity. Yet all the definitional questions about human rights raised by Hannah Arendt also apply when they are attached to energy and environment.[10] What would it mean to subject water and energy development to human rights? Who defends these rights and in what jurisdiction? Which usage rights trump others?

The very definition of human rights directly invokes the limits of political sovereignty. Whereas the former have relatively recent political traction in South America, the constitution of sovereignty and self-determination of the peoples have been sites of contention dating to the Conquest. Sovereignty doctrine, the fundamental precept of international law (including treaties and economic agreements with multinational corporations and therefore the legal apparatus surrounding Itaipú), assumes that all national states are equal and equally sovereign within bounded territories. Yet, in the international order since the sixteenth-century European-American encounter, the sovereignty of non-Western peoples and of smaller countries has been repeatedly subordinated to countries in the Global North, to international financial organizations like the IMF, and to multinational corporations. That is to say, not all sovereigns are equally sovereign.

By universalizing and totalizing Western values and conceptions of jurisprudence over and above indigenous legal practices, international law provides a mechanism by which to dispossess peoples of territory and resources: sovereignty doctrine has prioritized certain cultural values—those of commerce and extraction—over others in the past.[11] This is one motivator for the legal recognition of rivers, mountains, and nature as persons meriting legal protection, creatively using the idiom of Western law to elevate non-Western cultural norms. Yet, given the track record from Standing Rock (North Dakota) to Yasuni National Park (Ecuador) to rare earth metal mining in Baotou (China), it remains to be seen whether human rights can offer a meaningful check against extractivism. The placement of sovereignty and self-determination as Item 1, immediately followed by thirty points detailing integration, illustrates an identified pattern where a weaker party's sovereignty is linked to and constrained by the rights of others to trade.[12] In a more extreme circumstance, international law may be used to dispossess people of control over their natural resources if they or their usages of the resources are deemed unfit, or if a community should be judged corrupt or naive stewards, or if a "greater need" for the resources will rise to supersede local needs, and so on. But in its milder incarnations, advances in law still have a system-stabilizing effect. For example, though the electricity-related provisions were advantageous to Paraguay (potentially eliminating the physical and, in the future, legal barriers to unrestricted access to its electricity), the price for peace was the acceptance of the construction debt, which stood as a proxy for a larger set of principles. Renunciation of the debt would not only leave a $30 billion deficit in Eletrobras; it would also set a precedent that would undermine established principles of economics and contracts.

While articulating a commitment to "diminishing poverty, inequality, and other forms of social injustice," the subsequent terms of the declaration look to economic development as the engine for progress and maintain the nation-state as the individual, independent unit that wields sovereignty. In addition to matters of energy, the text also treats other bilateral issues of importance for Brazil and Paraguay: (a) a Unified Tax Regime (UTR) linking Paraguay's Ciudad del Este to the Brazilian tax system and (b) the large "brasiguayo" population of Brazilians living in Paraguay. Economic interests around international commerce (in the UTR) and international agribusiness (the employment of many Brazilians resident in Paraguay) were raised and resolved in the Joint Declaration. The triad of energy, agribusiness, and commerce points to what is being integrated, suggesting that the integration that is forming the region might not be so much the fusing of individual nation-states by the removal of borderlines and passport requirements, as it is the melding of transnational commerce, transnational agribusiness, and transnational energy, now endowed with millennial hopes of social justice. Nonetheless, the opening items (effectively the preamble) do not mention economics or financial rights. And the threads held together by the Joint Declaration may be pulled in divergent directions—human rights, self-determination, solidarity, social justice—that is, those elements that prioritize wide human communities versus the mechanics of those goals, which are integration and economics.

Energy Integration

Even before addressing the issues of disagreement within Itaipú, the Joint Declaration expands on energy integration, subordinating the dam to the political economy of energy in the region with Item 4 saying, "[The presidents] reiterated their commitment to the energy integration of the region and underlined its potential to advance social and economic development and the eradication of poverty" (Joint Declaration 2009). For all their differences, the Hydroelectric Commission and the Itaipú technocrats, too, strongly advocated for integration. It is a curious thing that the political economics of energy in the region should be idealized as "energy integration." In other emerging markets, a kind of "resource nationalism" has increasingly characterized the use of energy and natural resources, particularly hydrocarbons. Resource nationalism refers both to the process of nationalization of ownership and control over resources, but also to the ways that governments use a political/strategic calculus (and not just an economic one) when making decisions about how to use these resources.[13]

Political scientists like Flynt Leverett point to the case of Bolivia's nationalization of control over its natural gas (wrested from Brazil) as well as Evo Morales's use of the proceeds from these sales as an instance of resource nationalism.[14] Moreover, one might signal the same in Fernando Lugo's electoral campaign promise to recover hydroelectric sovereignty. Yet Lugo's energy experts coincided with Brazilian negotiators to proclaim energy integration as the cure for social and political ills. The Joint Declaration's energy-related provisions enact a careful balance, attempting to settle the contention between Paraguay and Brazil as well as outlining a path for a regional future. Though it might seem that resource nationalisms rather than integration should be on the rise in the midst of Anthropocene pressures, there is something in the particular nature of hydroelectricity and water resources in general in the region that lends itself to integration. In fact, this is necessary given the vastly uneven distribution of the production and of the consumption of electricity in South America—the former a result of the geography of natural resources, the latter of the geography of industrial development.

The declaration's inclusion of energy integration was part of a larger regional trend toward energy integration, which rests on twinned infrastructures, one legal, one material. The proposed Union of South American Nations (USAN in English; UNASUR in Spanish) treaty on energy integration accompanies a growing lattice of natural gas pipelines and high-tension electric wires that stretch across national boundaries, of which Itaipú is but one node (albeit crucial). In 2008, ministers and diplomatic staff from Spanish-speaking, Portuguese-speaking, English-speaking, and even Dutch-speaking countries in South America and the Caribbean met in Brasilia to discuss political, economic, and social challenges facing their respective countries. However, the main topic at hand was energy integration. Debates had gone on for years, but in 2008, largely at the urging of Lula's government in Brazil, they decided to take the matter seriously. As the delegates sat to compose a plan for energy integration, they instead drafted a treaty proposing an economic, social, and political union of South America. The Union of South American Nations took smaller regional blocs—Mercosur (Argentina, Brazil, Paraguay, Uruguay, provisionally Venezuela), the Andean Community (Bolivia, Colombia, Ecuador, Peru)—and integrated them. The union affirmed but nested over the smaller multinational agreements, meaning that Mercosur and thus Itaipú are legal antecedents for the union.

The union, now in effect in the continent though heavily embattled, is a combination of something like the EU and NATO.[15] At the height of its popularity, members proposed forming a coordinated South American Defense Council, a South American Health Council, and a South American Development Bank. While USAN was formally adopted, the Energy Integration Treaty was merely tentative. The official planning document "USAN: A Space That Consolidates Energy Integration" highlights the proposed structure of the treaty.[16] Energy integration calls for deeper infrastructural linkages between countries, the complications of which the Itaipú case illustrates (requiring a harmonization of electrical grids across varying hertz and the upstream/downstream adjudication). Moreover, integration requires the coordination of energy policy between countries with bellicose pasts (Argentina/Brazil, Argentina/Brazil/Paraguay/Uruguay, Bolivia/Peru/Chile, Paraguay/Bolivia, Ecuador/Peru) and yet without a shared currency. Synchronizing supply and demand across national boundaries, also called for by the proposal, requires complex arrangements of temporality; hydroelectricity has different rhythms of production-consumption than fossil fuels.

As with the Joint Declaration, in addition to stipulations regarding energy and environment and respecting the territorial integrity and full sovereignty of various nation-states, the proposed USAN energy integration treaty acknowledges preestablished guiding principles found in previous declarations that include important statements regarding rights. The authors frame the endeavor in similar liberal democratic terms as the Joint Declaration by stating as basic principles "solidarity among the peoples" and "respect for sovereignty and self-determination of the peoples," though they do not explicitly name human rights.[17] Among the goals of energy integration are (1) energy security, especially in the context of "global economic crisis" and in the context of South America's "privileged status" vis-à-vis natural resources; (2) the "benefits of a modern lifestyle" for all citizens; (3) the recognition of "energy access as a citizenship right."[18] Because rights are connected to citizenship, the specification of "modern lifestyle" and "energy access" as inalienable components of national belonging refines the energy dimensions of political recognition. If access to resources indexes identity, then the USAN Energy Integration Treaty argues that all people in South America, not just elites, should have access to energy and economic development. In the world region with the greatest levels of inequality, this has emancipatory potential and an implied social justice critique.

Energy integration thus extends and fulfills the promises of economic and political liberalism. The ecoterritorial turn we see in energy integration contends with lacunae of Western political-economic governance that have become untenable: either subordinating nature to culture or ignoring it entirely. To the degree that government increasingly takes on ecological dimensions, in the language of energy access as a citizenship right we see the seeds of an ecocitizenship—rights and responsibilities between the human community and the state apparatus that governs it. In lieu of monetary circulations that harmonize a region and provide sensuous evidence through which a community may be imagined, the physical infrastructure of high-tension wires, hydrocarbon pipelines, canals, and railroads, each moving with their own rhythms and to be coordinated by USAN, may offer a scale-producing alternative circuit. Perhaps, following the parallels between current, currency, and debt in chapter 4, energy itself will be the ecomoney that circulates in this regional state, flowing across wires or through pipelines that carry oil and natural gas.

Conclusion

Ecoterritorial governance works by leveraging significant resources to effect substantive changes in national and/or regional political economics. In South America, energy integration is a spatial fix to the limits of capitalism that works by reorganizing governance and redistributing risk along new scalar contours. Just as Itaipú and the Joint Declaration were a dress rehearsal for the Guaraní Aquifer, the USAN attempt at an Energy Integration Treaty serves as practice for how to deal with what is arguably the most important source of freshwater in the world, the Amazon. Two human-created infrastructures (one of law in USAN and the Joint Declaration, one in the physical high-tension lines and pipelines) form the nexus around which a South American regional state-like apparatus may coalesce, agglomerating governments, energy companies, built environment objects, rivers, watersheds. The lines provide a territorial basis, the treaty agreements a governing apparatus; and all the citizens of the member countries provide the human community. And though there are many aspects to these changes, an attention to the jurisgenerative character of treaties looks at legal imagination and the political modalities necessary to effect the ecoterritorial turn. At its most basic level, energy integration solves the temporal problem of natural resource scarcity by flexibilizing the sources of energy. The documents also create legal flexibility. Neither the Joint Declaration nor the Energy Integration Treaty mentions rights of nature; rather they focus on the human and consumption rights of people, and, since treaties supersede national legislation, these two documents provide a counterposition to Ecuador's and Bolivia's laws.

While no particular nation-state can claim water resources that evade national boundaries, a regional state apparatus can. The Union of South American Nations and the USAN Energy Integration Treaty fully encompass and encircle the natural resources of both the Guaraní Aquifer and the Amazon. Regardless of the sketchy definition of region, with the massive water resources of South America in mind, the region also becomes a way to counterbalance the global. It asserts local rights to water. And the more expansive legislation around natural resources, the linking of rights and social values to energy, bestows legitimacy to local administration of water in South America. In the sixteenth century, a group's rights to administer natural resources could be removed when they prevented trade. In the twenty-first, when other, larger, more powerful countries have need of water and when natural resources are imagined as a global good under the rubric of sustainability and not merely a local good, a thickened chain of international legal apparatus redistributes the weight of international law in favor of the South American nations.

The force of treaty law is not just its ability, in a logocentric fashion, to script reality into being. Law points to power structures that accrete authority—legal precedent, political ascension, financial interest, educational credentials—performing a skillful balancing act by including a motley of provisions urged by powerful elite sectors as well as following along legal principles that were hard-fought popular struggles. Because of the work expended in creating and then signing the document and because of the platform in which it was signed, treaty law sets precedent: exalting government as the solution to social crisis; explicitly linking the economic futures of energy, commerce, and agribusiness; tying social justice to economic development; binding integration, sovereignty, and human rights. But regional integration meets its match in the value of sovereignty. And because human rights are a universalized value, they can legally check the economic imperative. This is where the emergent value of sustainability offers new potentials in twenty-first-century international law. Like human rights, sustainability is a universalized value, appealing beyond the boundaries of the nation-state (or a region). Claims to safeguard sustainability, therefore, may be a way to check and supersede the limits placed by human rights.

The Joint Declaration and the USAN Energy Integration Treaty also demonstrate the political economics of energy organized on a level that transcends national boundaries and fluctuates between different scalar dimensions—at times on a nation, at times on a region of fused-together nation-states, at times on projections that are watershed determined. Such fluctuations might be interpreted to mean that we are witnessing a transition from one scale of organization to another, that the uncertainty and lack of fixity is a stage in a process by which political economic formations are reconstituted on new scales. However, my intuition is that the political economics of energy shifts scale in a different way—that it vibrates on a number of harmonics at the same time and that this multiscalar appearance is not a transitional phase, but the way it works. And so this is what "jumping scale" means: not that the political economics of energy switches up (or down), but that new ranges are included as when multiple chords are played at the same time (and that the converse may be true, that harmonic ranges are silenced).[19] So, the national scale is important in the organization of human labor and the means of production (including natural resources), but, in terms of circulation and consumption, the natural resource base of energy mandates something beyond national boundaries. The uneven development between national units in South America (where some countries have higher levels of consumption than others, where even the natural resources are distributed unequally) couples with an "international" (transnational?) imperative. The Joint Declaration and USAN do not nullify the nation-state as a form of organization; instead, they further the development of an additional form of organization. This is not a step toward the future obsolescence of the nation-state en route to "globalization" but solidifies both the region and the nation-state as simultaneous, sometimes integrated, sometimes harmonious, sometimes discordant.

Conclusion

The question of the future looms large for Itaipú Binational Dam. In 2023, important parts of the binational treaty must be renegotiated, and the financial arrangements are once again up for grabs.[1] That year, unless an additional US$4 billion in debt is discovered, as occurred with Yacyretá Dam (Argentina-Paraguay) in 2014, the massive construction debt will be paid off. But it is unclear how this will affect the tariff. If the present tariff formula holds and energy production costs are the limits for the energy price, the tariff will plummet by more than 60 percent. Paraguayan industry and the greater public favor this, but energy managers and the Paraguayan government are more reticent. Energy experts are concerned that this will lead to haphazard energy waste by the largest consumers, rather than a strategic industrialization plan. On the other hand, politicians in the Paraguayan executive and legislative have begun discussions on what to do with US$1 billion more in royalties—if current electricity sales come to nearly US$4 billion and about 60 percent of that (more than US$2 billion) goes to debt, then US$1 billion would be freed up for both the Brazilian and the Paraguayan governments. But the hydrostate history of Itaipú illustrates the weaknesses in either transferring more energy rent to the Paraguayan government without enforceable development investment plans or of effectively subsidizing Paraguay's current electricity users, who are disproportionately wealthy, without implementing a strategy to increase energy access to the vast majority of the country.

Itaipú 2023 requires deft arbitration. The track record of the past has been that "the solution for Itaipú is within Itaipú," suggesting that presidentially appointed Itaipú managers in Paraguay, not popularly elected representatives or trained career diplomats, may have a commanding role in one of the most important energy negotiations of the early twenty-first century. The new tariff negotiation occupies a central place in Paraguayan public discourse. There is even talk of having economist Jeffrey Sachs take a leading part in treating on behalf of the Paraguayan government. But other deadlines and energy scenarios are more muted.

Paraguay has experienced a soy-fueled economic boom over the past two decades and benefitted from the market instability of its neighbors in recent years as Argentine and Brazilian companies move offices or investments to the smaller country to shield them from political and economic volatility. Income gains and the completion of the five-hundred-kilovolt transmission line between Itaipú and Asunción making more electricity available in the growing capital region have led to rising rates of electricity consumption, which have grown between 8 and 12 percent a year since 2010. When I first visited Itaipú in 2007, Paraguay consumed 5.6 percent of the energy of the dam to meet about 90 percent of its entire national demand.[2] In 2016, Paraguay consumed 11 percent of the dam's record-breaking production to satisfy only 75.6 percent of the country's demand.[3] That is, Paraguayan consumption from Itaipú has doubled in only a decade. As early as 2023, ANDE has predicted, but more likely closer to 2035, Paraguay's electrical demand will require all its installed capacity in Itaipú, Yacyretá, and Acaray Dams.[4] At that point, compensation payments would cease, disrupting the municipal and departmental governments who acquire more and more of their annual budgets from hydroelectric rent. And should Paraguay require more electricity, the only viable sources would be to purchase Brazilian energy from Itaipú and/or Argentine energy from Yacyretá.[5] Paraguay might find itself *paying* compensation to the victors of the War of the Triple Alliance—a fate that neither the media nor politicians have mentioned in public.

Nevertheless, the new Itaipú negotiations offer the opportunity to rethink more than just royalties or compensation, but the very value of water in a time of ecological crisis and climate change. Electricity is a principal way developed countries consume energy. For example, in the United States, nearly the same amount of fossil fuels is destined for electricity generation as goes to transportation.[6] Electricity, therefore, is not a sideshow in energy production; it is the main event. And Itaipú Binational Dam is the world's single largest power plant—hydrocarbon burning or otherwise—in terms of electricity generated. Because energy is a form of cultural production, it offers a way to see how social values embed in and emanate from the material world, from nonhuman living beings, and from organic and inorganic things. Sociocultural anthropologists study the meaning-making systems within which and with which human beings live; environmental anthropology connects cultural values and structures to ecosystems, climate, and geology in order to explore the relationships human communities have with the material world around them. And so, this book on Itaipú hydroelectricity has traced the political and economic systems that arise from an electrification predicated on water, not fossil fuels, to show how the material basis of energy, although electricity may be effectively fungible, nevertheless marks the infrastructure, the politics, and the economics animated by that energy source.

Itaipú rests in a flow, not just of hydraulics or charge or money, but of political-legal thought and affective relationships to water as law manifests power. The dam, both product and producer of regional changes, is entangled in a complex that emerges around natural resources, community belonging, and the future. I have argued that, through Itaipú and in the region more broadly, hydropolitical dynamics centered around the values of sovereignty and security dominate energy politics. Thus, a political economy with renewable energy contours favors (a) strong state involvement in economic development; (b) state control of natural resources with a view toward environmental sustainability; (c) the use of engineering to resolve political dilemmas such that science is a form of politics as a vocation; (d) supranational cooperation and integration that leads to the formation of a regional state-like apparatus; and (e) a rhetoric and politics of paternalistic state stewardship of nature for the sake of the national community, often traversing and reshuffling nation-state boundaries. With the challenges of anthropogenic climate change, alternative energy sources become mainstream. As we have seen, the contest of values is rooted in real-life politics, with personalities and partisan histories playing a major role. This was one difficulty with Paraguay's negotiating attempts and the development futures of chapters 3 and 5. If the physical infrastructure of Itaipú smoothed competing political agendas and the financial architecture harmonized disparate accumulation strategies, the legal apparatus around Itaipú synthesized national and regional goals.

In Itaipú, hydroelectricity, scientific industry, and people themselves are objects of engineering interventions. The destruction of Guairá Falls was justified not just by the ever-expanding demand for electricity in Paraguay and Brazil, but by the ever-increasing efficiency of the turbines in Itaipú and the regular record-breaking energy production. Technical excellence atoned for the sacrifice of nature. From the very beginning of the international diplomatic negotiations in the 1960s to the new crises nascent in the Lugo administration in the twenty-first century, technical expertise was central to governance. At the helm of Itaipú decision making, of negotiating the requisite international diplomacy and drafting of national energy policies, of designing the economic strategy and implementing social-development projects were the engineers themselves. The contests over the ideal nature of state-to-nation relations, over how to manage the economy and resources, and over how to achieve progress strike to the very heart of what kind of future is being imagined through Itaipú electricity.

The first two chapters discussed interlaced circulations, showing the parallels between current and currency. In chapter 1, we saw how the physical infrastructure of the hydroelectric dam produced megawatt hours of electricity and protected national sovereignty. Recognizing that power is politics, the engineers have steered the legal and the financial, and not merely the technical, and have been instrumental in shaping the ethics and governance of renewable energy. Chapter 2 showed how the complicated financial infrastructure for the binational dam bore the imprint of the financial logic of the dam—economic equilibrium and environmental sustainability—resolving into the tariff, the price for electricity. The flow of money into and out of Itaipú was explained by the "making of a tariff" as one single number hid complex calculations and obligations and compromises. As part of the hydrodollar effects of Itaipú energy, the tariff linked three different accumulation strategies: monopoly rent (the tactic Paraguayan elites favored), a market-based price versus one set by fiat, and compound interest (the strategy benefitting Eletrobras and its shareholders). Because Eletrobras was responsible for compensating the Paraguayan side for Paraguayan energy sold in Brazil, it also paid a compensation that was configured as a non-market-rate moral debt stemming from the War of the Triple Alliance.

Itaipú transforms circulations into integrations. The presidential election of former bishop Fernando Lugo in 2008 initiated a new round of Itaipú negotiations as the Paraguayan government sought to shift the power- and revenue-sharing relationship with Brazil in the dam. Chapter 3 traced the strategies of two groups, the Ministry of Foreign Relations Commission on Binational Hydroelectric Entities and the Itaipú Paraguayan Executive Directorate, as they leveraged the dam to secure larger political futures for the country. The chapter juxtaposed public performance and extended personal interviews conducted with many of the participants to elucidate what occurred behind closed doors. As part of sweeping political changes across the continent, the internationalisms practiced by the two energy groups depended on how each imagined the state, the locus of state power in Brazil, the relationship of the nation-state to the broader international state system. The Hydroelectric Commission focused on mobilizing the nation, while the Itaipú technocratic leadership strategy engaged Brazilian political and economic leaders.

Eletrobras benefitted from a conflict of interest in setting the Itaipú tariff too low to meet the interest rate of its loans to the dam from 1984 to 1996, a story narrated from several vantage points in chapter 4 as a crucial effect of an asymmetrical integration. Whereas Itaipú had only two shareholders—ANDE and Eletrobras—the Brazilian utility company had many because in 1995, in a flurry of neoliberal privatization throughout the region, a minority stake in Eletrobras was sold to private investors. Yet the largest dam in the world circumvented the expansion of new market orthodoxies (i.e., Itaipú was never privatized). Even now, individual shareholders of the publicly traded company reap benefits from the politically anchored financial decisions in Itaipú. But perhaps as telling for the future is the historical role of credit in financial systems. Debt lashed Brazil and Paraguay together, suggesting that energy itself may become a currency of the Anthropocene.

Through Itaipú, water is converted to policy and international law as a way to secure the future. Victory, albeit limited, appeared in the form of the Joint Declaration, signed by Paraguayan president Lugo and Brazilian president Lula da Silva in July 2009. With the promise of funding, the leaders in the Hydroelectric Commission and the technocrats in the Itaipú Executive Directorate discussed development futures, as we saw in chapter 5. Although both the commission and the directorate agreed that hydroelectric-led development would be the cure to Paraguay's underdevelopment ills, conflicts abounded over the definition of good government. Lugo's progressive base used participatory democracy to describe the political project that the new government would implement through spending focused on social development. On the other hand, the technocratic team in Itaipú envisioned development in terms of economic growth, educational attainment, and the strengthening of democratic institutions, directly tied to the industrial potential of Itaipú Binational Dam. Partisan distrust and political rivalries dissolved the Lugo coalition and stalled nearly all the Joint Declaration's measures (and the development plans of both Canese's and Balmelli's teams) until only the energy rent gains remained. And then Lugo was impeached, leaving government agencies in the hands of the Liberal Party for one year until the new presidential elections, when the Colorado Party returned to power.

As chapter 6's discussion of legal-intellectual frameworks around energy shows, the Joint Declaration and the Energy Integration treaty solidify both the region and the nation-state as simultaneous, wherein the region also acts as an important anticipatory countermeasure to the global, particularly when natural resources are concerned. International legal apparatus around transborder water and energy sources increasingly assert that the region, or, as the Parlasur representative from Uruguay said, "this South America," is the proper adjudicator for resources that have global significance: the Guaraní Aquifer and the Amazon. The natural-resource-administering regional state, geographically larger than the individual nation-states of South America, is simultaneously activated by something unimaginably small: the electron-scaled circulation of national electricity (chapter 1).

In the Anthropocene, the sense of future imminent doom, almost inevitable and fated, bends back on the present as we realize that humans have become a world-altering force. Drought caused by anthropogenic climate change has put renewable energy resources in jeopardy; and 2014 saw rolling blackouts and shockingly low reservoirs in Brazil. But then 2016 broke world records of energy production again. Core to the value-added of hydroelectricity is its renewability because of the fact that energy extraction does not, in theory, diminish the resource. Nature refills the river. It is thus a sustainable source of energy. But another key part of hydroelectricity's value in the region is political-economic: prominent reliance on hydroelectricity, much of which is fully or partially state owned, subordinates industry to state actors, leading to hydropolitics.

Other impending crises have ever-nearing deadlines. By 2100, the dam will approach its mechanical obsolescence, leaving Brazil and Paraguay with an energy deficit of fourteen thousand megawatts not easily replaced because of the unique hydroelectric potential of the Guairá Falls. The security and sustainability of state power and industrial profits are also jeopardized by anthropogenic climate change. In the twentieth century, violence in defense of state sovereignty was associated with renewable energy. I wonder if we should pay attention, in the twenty-first century, to how engineering decisions in the Anthropocene framed as rational responses to natural disasters, in the name of sustainability, result in violence and oppression. The ethics and the political economy of renewable energy in Itaipú Dam complicate pressing conversations about energy (and water use) taking place in and outside academia. Yet, because Itaipú trades not just in energy, but in hope, hydroelectricity has also been a story of how water unites.

NOTES

Introduction

1. For a wide-ranging discussion on sovereignty, international legal sovereignty, and necropolitical sovereignty, see Agamben 2005; Anghie 1999; Arendt 1990; Krasner 1999; Hansen and Stepputat 2006; Schmitt 1985; Williams 1990.

2. Recent Paraguayan administrations have, in fact, taken Brazil up on its offer regarding relying on water flow management from the upriver dams. Doing so makes electricity production more efficient because it reduces uncertainty.

3. The notion of a river as a "machine" comes from Richard White's (1996) analysis of the Columbia River, which he configures as an "organic machine," whose physical configuration coincides with shifts in energy extraction, from salmon to hydroelectricity.

4. Energy Information Administration 2017a.

5. Energy Information Administration 2017a; Ministerio Coordinador de Sectores Estratégicos n.d.; Organismo Supervisor de la Inversión en Energía y Minería n.d.; Telesur 2016; Unidad de Planificación Minero Energética n.d.

6. Wittfogel 1957.

7. Aravamudan 2013; Chakrabarty 2009:212; Crutzen 2002:23; IPCC 2014.

8. Leslie White's 1943 article reveals long-standing anthropological concerns regarding cultural ecology and, specifically, energy.

9. Morgan 1877.

10. Love and Isenhour 2016; Sawyer 2004; Tsing 2005.

11. Kraay and Whigham 2005; Lambert 2006.

12. Pronounced "why-rá" (rolled "r"). Curiously, the Spanish language version of this treaty differs from the Portuguese in a small, but significant, way. The Spanish says "El territorio del Imperio de Brasil se divide con el de la República del Paraguay por el cauce o canal del Río Paraná," while the Portuguese says "O território do Império do Brasil divide-se com a República do Paraguay pelo álveo do rio Paraná," As far as I can tell, this has not yet become a source of conflict, but it does imply the border is fixed in Portuguese and more flexible in Spanish.

13. Stroessner 1965.

14. Blanco Sánchez 1968.

15. J. H. White 2010.

16. California Energy Commission 2016:1.

17. The rest comes primarily from other hydroelectric dams and secondarily from biodiesel and/or fossil-fueled thermal plants (i.e., from ethanol-burning or fossil-fuel-burning steam plants).

18. The effects and affects of [state] violence are explored historically and ethnographically in Abrams 1988; Benjamin 1986; Coronil and Skurski 2006.

19. Folch 2015.

20. While Anglophone anthropology has paid much attention to states of exception and the discussion on sovereign power drawing from Schmitt (especially following the attacks of Sep-

tember 11 and the subsequent use of rendition and of Guantanamo Bay as a detention center for enemy combatants, and the suspension of habeas corpus via the Patriot Act), an older debate on sovereignty runs through international law, based on the Spanish–New World encounter and the question of the sovereignty of indigenous communities in the Americas; see Anghie 1999; Folch 2016; Kennedy 1986; Ramón Hernández 1991; Scott 1934; Vitoria 1917 [1532]; Williams 1990.

21. German electrical manufacturer AEG acquired a concession in 1897 to electrify Buenos Aires. American and Foreign Power Company, with holdings in Cuba, Guatemala, and Panama, also moved into Argentina in 1926, where it remained until the 1960s. With more suitable coal deposits, Argentina developed thermal generation plants such that, by the time electricity was nationalized under Perón, the hydro sector had never become the backbone of electricity generation that it did in Brazil. Today, hydropower provides 30 percent of Argentina's electricity, an unusually low number compared to the rest of the region.

22. ANDE n.d.

23. Cattelino 2008; Folch 2015, 2016.

24. Bulmer-Thomas 1994.

25. Fogel and Riquelme 2005; Hetherington 2011, 2014; Pastore 1949; Weisskoff 1992. Paraguay's rapidly expanding soy frontier has largely been developed by Brazilians, known as Brasiguayos for their residence in Paraguay. Paraguay is the world's fourth-largest soy exporter, behind its neighbors Argentina and Brazil as well as the United States. Recent struggles around soy in Paraguay include the aim of progressive sectors to raise the soy export tax to 10 percent, an initiative that failed.

26. To placate the population and foster a citizenry that supported the Colorado Party government, Alfredo Stroessner's military government (1954–89) also initiated an agrarian reform to distribute unused land to Paraguay's campesino population, though much of the land instead went to Colorado Party and military elites (Kleinpenning and Zoomers 1990). As part of a growth of Inter-American Development Bank and foreign-funded infrastructure (roads, bridges, electrical lines), Paraguay also took advantage of the take-off of its eastern neighbor via border arbitrage and, of course, Itaipú (Nickson 1982; Schuster 2015).

27. Hetherington 2011:12.

28. During the war years, the balance of trade was in Brazil's favor, but once peace returned, agribusiness prices (e.g., coffee) dropped, and import costs soared, leaving the economy in a lurch. Under the direction of state agencies and sheltered by protective import tariffs, metallurgy, machinery, and manufacturing grew (Colistete 2010).

29. See Hetherington and Campbell 2014 for more on infrastructure as a modernizing project in South America.

30. Though associated with Pinochet in Chile in the 1970s and Thatcher and Reagan in the 1980s, in the 1990s and 2000s the economic philosophy spread through the rest of South America. Neoliberalism was framed as a return to liberal economic orthodoxy designed to halt inflation and bring economic growth by removing the state (deemed inefficient) from the market (deemed a more efficient way of distributing resources). The formula included the cutback of government welfare programs, the privatization of state-owned industry, and the growth of private providers of heretofore state-provided goods (e.g., health care, water, and education). See D. Harvey 2007.

31. Ong 2006.

32. Gudynas 2011.

33. Security concerns centered in Ciudad del Este, particularly around Islamic terrorism, rose to prominence following the attacks of September 11; see Bartolomé 2002; BBC Mundo 2005; Hudson 2003. Nevertheless, recent historical ethnography of the region has nuanced and troubled these depictions (Karam n.d.).

34. *ABC Color* 2009a.

35. Folch 2013.

36. Timothy Mitchell (2011) has looked at the labor arrangements, extraction techniques, and transportation of fossil fuels to argue that coal and petroleum coincided with a shift from strong national labor movements that checked elite forces to strong transnational capital and weakened labor.

37. I follow Michel Rolph Trouillot's call to study state-like effects regardless of whether they emerge from national governments or other polities (Trouillot 2003:89). Trouillot expands on Mitchell's (1999) use of the term to draw attention to the state as a series of effects as a way to avoid reifying the state as an identifiable entity separately distinguishable from "society."

38. Coronil 1997; Mallon 1995; Mintz 1986; Nash 1993; Sawyer 2004; E. Wolf 1982.

39. Bayart 2000.

40. Nader 2010.

41. Masco 2006.

42. Drackl̩é and Krauss 2011; McCandless 2007; Tulachan 2008.

43. GISE (Grupo de Investigación en Sistemas Energéticos) of the Universidad Nacional de Asunción–Facultad Politécnica is conducting groundbreaking interdisciplinary energy research in Paraguay; see Blanco et al. 2017. Dam building in Africa (Kenya/Rukenya Dam), in Central America (Costa Rica/PH Diquís Project), and on the Mekong River (Laos/Xayaburi Dam) elevates certain priorities in water management over others. For example, while water supply might be more urgent in Kenya's Rukenya Dam, in the Costa Rican dam, power generation is more important. What we are seeing in the Mekong is that industry in neighboring Laos trumps the livelihoods of the millions who live off the river in Cambodia and Thailand.

44. Amery 2002; Poff et al. 2003; A. Wolf 1999.

45. Anthropologists have a long-standing role in analyzing the unintended consequences of dam building on local communities (Cernea 2004; Colson 1971; Dorcey 1997; Johnston 2013), including excellent ethnography during the construction of Argentinian-Paraguayan binational dam Yacyretá (Ferradás 1998; Lins Ribeiro 1994). Curiously, the same attention was not given to Itaipú.

46. Petersen 2014.

47. Folch 2013. See also Blanc (2015) and J. H. White (2010) for the labor history and social impacts of land expropriation during Itaipú construction.

48. Chibnik 2011.

49. Blanc 2017.

50. Douglas and Wildavsky 1983; Masco 2014; Orlove 2002.

51. Empresa de Pesquisa Energética 2017:16.

52. Academia Brasileira de Ciências 2015.

53. Energy Information Administration 2014.

54. Ventura 2018.

55. Mauss 2000:78.

56. Mauss 2000:79.

Chapter 1. Current

1. MCDyA 1977.

2. MCDyA 1977.

3. "Energy" is the capacity to do "work"—to change temperature, move something, or illuminate.

4. Latour 1986 [1979].

5. It also hearkens to Wittfogel (1957), who found in rain-based and irrigation-based agricul-

tures very different political-economic structures, the former much more locally independent and the latter much more centrally controlled.

6. Weber 2004.

7. Mukerji 2009:216.

8. The study of infrastructure is enjoying renewed popularity among the social sciences—in part because it allows researchers to theorize relations between the material, aesthetic, and social. Susan Leigh Star's 1999 article was instrumental in calling ethnographers to the task, particularly by insisting on attention to "the values and ethical principles" inscribed in information systems (Star 1999:379). See also Larkin 2008; Graham and Marvin 2001; Schwenkel 2015.

9. Because infrastructure points our attention to systems, within the last two decades, the heuristic has become more popular within anthropological studies of a broad range of phenomena, including language, media, money, state regulation, and, of course, the administration of natural resources; see Appel 2012; Fischer 2005; von Schnitzler 2008, 2013. In this, anthropology has taken a geographical turn, but one in keeping with the corrective Marxist geographer David Harvey has urged in keeping the temporal equal to the spatial (D. Harvey 2006).

10. Although not always in use today, the spillway operated constantly for a decade (1982–92) while the Itaipú turbines were installed.

11. Hydrometeorological Telemetry Systems (HTS) sensors placed along Brazilian waterways relay river and rainfall data automatically to the dam. A human element also comes in. Station staff located at ten upriver monitor stations owned by Itaipú and three belonging to the Brazilian National Electricity Agency (ANEEL) regularly telephone or radio information to the dam's control room.

12. This comparison comes from Itaipú itself: https://www.itaipu.gov.br/en/energy/generating-units, accessed May 27, 2014.

13. The United States uses British thermal units (BTUs) in place of watt hours; 1 BTU is 0.293 watt hours.

14. California Energy Commission 2016:1.

15. Itaipu Binacional 2017:25.

16. Acuerdo Tripartito (Argentina-Brasil-Paraguay) 1979. Of the actual energy available, 9,637 megawatts (average, 2006–12) are available for Brazilian consumption; 863 megawatts (average, 2006–12) are available for Paraguay; or 76.5 percent compared to 6.8 percent (Itaipu Binacional 2013). Itaipú itself consumes the 17 percent remaining energy available.

17. Depetris 2007.

18. Marengo et al. 2012.

19. Winichakul 1997.

20. Mukerji 2009.

21. Mitchell 2011.

22. Folch 2013.

23. MCDyA 1977.

24. MCDyA 1977.

25. MCDyA 1977. For the sake of contrast, the Yacyretá Treaty (Argentina-Paraguay) has the physical installation of the dam as the condominium, meaning that there are no differentiated Argentine or Paraguayan turbines and that the border line is unclear.

26. Céspedes 1982:2.

27. Days later, government officials convened a special commission on the frequency issue including Debernardi, the foreign minister, and the minister of the interior (responsible for disciplining the internal opposition to the regime) as well as invited foreign experts.

28. MCDyA 1977.

29. MCDyA 1977.

30. Dinges 2004. The support or license of the United States in Operation Condor has long been suspected, though officially denied. Nevertheless, declassified documents reveal that secretary of state Henry Kissinger was informed in 1976 by Argentine foreign minister Admiral Cesar Buzzetti that Southern Cone countries were engaged in a joint effort to combat subversion. See Department of State 1976.

31. Folch 2013.

32. Folch 2018.

33. Jornal do Brasil 1977.

34. Jornal do Brasil 1977.

35. Folha de São Paulo 1977.

36. El Litoral 1977.

37. El Litoral 1977.

38. Itaipu Binacional 1994:2.34.

39. Inter-American Development Bank Project Number PR0002: Transmission and Expansion of Electric Generation System, http://www.iadb.org/en/projects/project-description-title,1303.html?id=PR0002, accessed November 30, 2016.

40. IACHR 1978: capítulo VII.

41. Folha de São Paulo 1977.

42. Mains 2012.

43. Folha de São Paulo 1977.

44. P. Harvey 2005.

45. Schivelbusch 1987.

46. *ABC Color* 2017.

Chapter 2. Currency

1. Bill Maurer (2012) calls attention to the "not-so-cool" aspects of finance, suggesting that debt has been considered "less cool" than credit and therefore studied less within the discipline. He writes, "We live in a world of capital but also of tribute and a world where rents themselves are declining in significance" (2012:480). The example of binational hydroelectricity in the Southern Cone, however, suggests that the world of rents is firmly entrenched in the domain of natural resources.

2. ANDE borrowed money, as arranged by Eletrobras, to provide this seed capital.

3. Wood 2016:45.

4. Coase 1937.

5. Coronil 1997.

6. Auty 1993; Karl 1997; Ross 1991. Terry Karl's well-cited *Paradox of Plenty* centers specifically on the example of Venezuela as petrostate, faulting weak institutionality (not the mere presences of lucrative natural resources), which she connects to imperial Spain's similar inability to manage the sudden influx of bullion.

7. A. Smith 2014:80.

8. Watts 2008:12. For ethnographies of oil states, see also Apter 2005; Limbert 2010.

9. Rogers 2014.

10. Plato 1967:304b.

11. Howe and Boyer 2015. See also Gupta 2015; Winther and Wilhite 2015.

12. Ballestero 2015:265.

13. Debernardi 1996:172. Several key figures in the administration of Itaipú have written accounts of their involvement in the dam, including the first executive director, Enzo Debernardi; the head of the Paraguayan engineering consortium and eventual Paraguayan president Juan Carlos Wasmosy (Wasmosy 2008); and executive director Carlos Mateo Balmelli (Mateo Bal-

melli 2011). Particularly the first two texts, containing financial reports and first-person accounts of historical events, provide important historical and numerical context to the dam.

14. Debernardi 1996:172.

15. Gregory 2012.

16. Itaipu Binacional 2018:45.

17. In Brazil, 45 percent to states, 45 percent to municipalities, 10 percent to the federal government.

18. Parlasur 2008.

19. Ho 2009.

20. Itaipu Binacional 2018:43. Under the new Brazilian government of former military officer Jair Bolsonaro (2019 - present), there have been rumblings within Eletrobras and the Brazilian energy sector in general to reverse this (Warth and Monteiro 2019).

21. *ABC Color* 2006. For more about Lugo's campaign, see Folch 2015.

22. Parlasur 2008.

23. See Espínola 2009.

24. Conclusion IV, Act of Foz do Iguaçu 1966.

25. Itaipú Treaty, article XV, paragraph 3. Literally: "el monto necesario para compensar a la Alta Parte Contratante que ceda energía a la otra."

26. Itaipú Treaty, annex "C," article III, paragraph 8.

27. The 1973 debate in the Paraguayan House of Deputies was heated and even drew public commentary, in spite of the tight control against popular criticism of the government from Stroessner's policing regime. "By buying even just one kilowatt, Paraguay is also paying the price of its own cession of energy, as does Brazil," suggested one Liberal Party deputy before the entire chamber, concerned that cession would be included in the general tariff (Gamón 2007:368). The deputy's comments mirrored critiques made in an open letter to the House of Deputies from the opposition Center for Engineering Students. "Finally, the compensation or earnings that Paraguay will receive as a result of the sale of its electrical energy to Brazil is included in the cost [i.e., tariff]. This point is absolutely inadmissible to Paraguayan interests," they wrote in 1973 (MCDyA 1977). The conversation in the house continued until one Colorado Party deputy flatly declared: "No one can cede energy to oneself. No one can, it's inconceivable, to oneself" (quoted in Debernardi 1996:281). And when the vote came, the treaty passed with the full support of Colorado Party deputies and most Liberal Party deputies (the "Radical Liberal" wing of the opposition voted nay).

28. Itaipu Binacional 2018:45.

29. Roitman 2004.

30. *ABC Color* 2006.

31. Canese 2007 [2006]:101. *La recuperación de la soberanía hidroeléctrica del Paraguay* (2007 [2006]) by Canese, has been republished several times and served as the manifesto that outlined the Lugo-campaign-cum-government's goal of recovering hydroelectric sovereignty.

32. Parlasur 2008.

33. *Búsqueda* 2009.

34. Itaipu Binacional 2008:61.

35. Flecha 1975.

36. CADEP 2012a.

37. CADEP 2012a.

38. Alcaraz Gavilán 2014:10.

39. CADEP 2010:14.

40. CADEP 2012b:20. These numbers also include Yacyretá revenue, which is minor compared to that of Itaipú. Also note, Codas was interim executive director for half of 2012, followed by an executive director appointed by the new Liberal Party Paraguayan president Federico

Franco. Part of the $237.2 million most likely also came from the newly implemented FONACIDE law; see chapter 5.

41. CADEP 2017:19, 20.

42. Appel 2012; Rogers 2012.

Chapter 3. Renegotiating Integration

1. A Tekojoja member was also in charge as director of the other hydroelectric binational dam, Yacyretá, at the time. Unlike with Itaipú, the top leadership positions in Yacyretá were never equalized, and there was a titular and organizational difference between the executive director (always Argentine) and the director (always Paraguayan) of the smaller binational dam. The expectation would thus be that, in his position, he would distribute "cargas de confianza"—appointed positions, determined by trust and the personal discretion of the administrator, rather than by tenure or competition.

2. Castañeda 2006.

3. Weyland 2003:1098.

4. Escobar 2010:7.

5. Mignolo 2005.

6. Das and Poole 2004; Evans et al. 1985; Sharma and Gupta 2006; Joseph and Nugent 1994; Nugent 1997; Tilly 1975, among others.

7. Mateo Balmelli even wrote an analysis of his time as executive director of Itaipú (Mateo Balmelli 2011).

8. Hetherington 2011.

9. Roa Bastos 1974. Roa Bastos fled into exile in 1947 during the civil unrest of the Higinio Morínigo government. He lived in Argentina and later France, returning to Paraguay in 1982 only to be expelled by the Stroessner regime for having criticized the administration.

10. Ministers of government in Paraguay, including the executive directors of Itaipú and Yacyretá, are referred to politely as Excellency. And Francia's contemporary, William Parish Robertson, writing of one of his last encounters with Francia before he and his brother were permanently banished from Paraguay in 1815, relates the following exchange:

> When I entered, I begged to congratulate "his Excellency;"—but here he stopped me short. "Déxe, amigo," said he, "de 'Excelencia,' y conozcame V. y hableme como hasta aqui hemos acostumbrado." "Lay aside 'Your Excellency,' my friend, and know me and address me as you have hitherto done." His title before was Usia (a contraction of Vuestra Señoria) which in Spain is a grade inferior than Excellency. But I knew the man I had to do with too well to avail myself of any such privileged familiarity. I continued, "Your Excellency," and he did not again object to the title. (Robertson and Robertson 1839: 17–18)

11. A new version of this list (one apparently not given to or accepted by Brazil's representatives) began to be circulated by the CEBH in late 2009, primarily to a Paraguayan audience. It renamed Point 1 "Soberanía hidroeléctrica" (Hydroelectric sovereignty) and Point 3 "Revisión y renunciación de la deuda espuria" (Revision and renunciation of the spurious debt)—more to the language used in Canese's 2007 [2006] book (pp. 101, 102).

12. Various explanations of why the meetings stopped exist: one version has the Brazilians abruptly ending the late January meeting after incredulously hearing that the Paraguayans considered renouncing the debt and the treaty; another version maintains that the negotiating teams were unable to come to consensus over whether to resolve the issue as a technical one or a political one—making it necessary to escalate the decision making to the presidencies. But whatever the real cause, what effectively occurred is that the Itaipú technocrat team took over all negotia-

tions. For the first six months of Lugo's presidency (August–January), there had been two simultaneous tracks of negotiations. Only one remained.

13. *Integración popular con soberanía energética y alimentaria construyendo la patria grande.*

14. Canese 2007 [2006]:101, 102.

15. *Pacta sunt servanda*: "Promises must be kept," meaning that treaties should be applied and fulfilled. *Rebus sic stantibus*: "Things thus standing," meaning that treaties may be inapplicable if fundamental changes have occurred that alter the initial circumstances under which the treaty was signed.

16. Juan Bautista Alberdi (1810–84), instrumental in writing the Argentine constitution (1853) and in establishing the political and social project of the nation—including the genocidal Conquest of the Desert and massive immigration from Europe to whiten the population of a barbaric nation—wrote during a time of intermittent war between Brazil and Argentina and is widely seen as an intellectual giant in South America. Alberdi's position against the War of the Triple Alliance has gained him respect in Paraguay, where he is seen as a key Argentine advocate against the devastating war.

17. Parlasur 2008.

18. Parlasur 2008.

19. Casa América 2008.

20. *Financial Times*, print edition, October 31, 2008, p. 9. See also *ABC Color* 2008.

21. Gualdani 2008.

22. See Gualdani 2008; Montero 2008; Peralta 2009.

23. Última Hora 2009b.

24. *ABC Color* 2009a.

25. *ABC Color* 2009b.

Chapter 4. Debt

1. Itaipu Binacional n.d.:15; Secretaria Técnica de Planificación de Desarrollo Económico y Social 2014:56; Belt et al. 2011:15. Each of these documents refers to unpublished ANDE figures. Taking the high, medium, and low end of Paraguay's electric growth rate from the last decade as base for calculations, a Duke University Itaipú research team (Duke Itaipú Post-2023 Team) estimated that at 12 percent growth, Paraguay will exceed its installed capacity in Itaipú by 2031; at 8 percent growth it will reach that threshold by 2037; and at 4 percent, by the more distant horizon of 2080 (Davenport, Folch, Vasu n.d.).

2. Contraloría 2012:19.

3. Parlasur 2008.

4. Brantlinger 1996; Graeber 2001.

5. Wilkis 2018:4.

6. Wilkis 2018:161.

7. Schuster 2015.

8. Smith-Nonini 2016.

9. D. Harvey 2007.

10. Graeber 2011.

11. Song 2009.

12. Gregory 2012.

13. World Bank n.d.a, n.d.b.

14. Debernardi 1996:160.

15. Debernardi 1996:160.

16. Craide 2009.

17. Gudeman 2001.

18. Deutsche Bank has a long history of financial investments in hydroelectricity. They partially funded the very first hydroelectric project, in Niagara Falls (Canada) in the late nineteenth century.

19. Itaipu Binacional 2003.

20. Contraloría 2012.

21. Baer and Beckerman 1989.

22. Contraloría 2012:25.

23. Canese 2007 [2006]:105.

24. Marx 1993.

25. Hetherington 2011.

26. González Echevarría 1990; Rama 1996.

27. Caplan 2000; Garsten and Hasselström 2003.

28. Declaración Conjunta 2009.

29. Bubandt 2009.

30. Peebles 2011.

31. Toledano and Maennling 2013:9.

32. Toledano and Maennling 2013:76.

33. Toledano and Maennling 2013:9.

34. Barbosa 2013.

35. N. Smith 1993, 2002.

Chapter 5. Neoextractivist Futures

1. Arendt 1990.

2. Jawarharlal Nehru, in the 1930s, also examined and wrote about the example of Lenin's electrification of Soviet Russia and its capacity to transform social life as a model for India on the cusp of postcolonial independence and modernity. Coleman 2017:92.

3. Biomass, obtained via massive deforestation to convert land for ranching and soy agriculture, provides industry with a significant fraction of the energy used.

4. World Bank n.d.b.

5. World Bank 2010:iii, iv.

6. Gudynas 2011; Svampa 2011.

7. The prominent role of the state in progressive neoextractivism, as proponent of natural resource extraction and then in redistribution of the proceeds, differentiates it from "classic" extractivism.

8. So-called because, after the massive natural gas discovery in 1959, the Dutch economy prioritized energy extraction to the neglect of other industrial sectors because of comparatively high profit margins. In the decades following the discovery, manufacturing declined. Dutch Disease also arose out of rising hydrocarbon pricing. As fossil fuels were traded globally for US dollars, local currencies of oil- and gas-rich states also strengthened in value, again making exports more expensive and increasing imports, with negative impacts on industry.

9. Other notable findings of this survey were the answers to two related questions regarding the ways responders perceived the present and previous use of royalties and compensation from Itaipú. The first question, "Are they used with transparency?" garnered the responses: 3 percent saying "a lot [of transparency]," 78 percent saying "little," and 16 percent saying "none" (with 3 percent not responding). The follow-up question, "How are the expenditures used?" received 16 percent "badly," 74 percent "normally," 5 percent "very well" (and 4 percent no response). That is to say, although 94 percent of Paraguayans surveyed said that the Itaipú funds had hitherto

been spent with little to no transparency, 79 percent of them said that the quality of the expenditures (the way the money had been spent) was normal or very good. Only 16 percent said they were badly spent. "Transparency" does not seem to be a fundamental criterion of good administration of public spending.

10. Hausman and Neuefeld 1997.

11. Barboza 1992.

12. Cardoso and Faletto 1979; Prebisch 1981.

13. Cardoso and Magnani 1974.

14. Última Hora 2009a.

15. Agencia Senado 2009.

16. Máximo 2009.

17. *ABC Color* 2009c.

18. Itaipu Binacional 2009b. The press release description of the event can be found at https://www.itaipu.gov.py/es/sala-de-prensa/noticia/mateo-balmelli-brindo-conferencia-en-curitiba.

19. The deal was also marred by popular dissatisfaction over the pricing arrangement, which committed Paraguay's energy for years; the few jobs created by the project; and rumors of multimillion-dollar bribes.

20. CADEP 2017:19, 20.

Chapter 6. Ecoterritorial Turns

1. Howe and Boyer 2015.

2. Parlasur 2008.

3. Folch 2016.

4. De la Cadena 2015; Viveiros de Castro 1998.

5. The 1966 Act of Foz do Iguaçu signed between Brazil and Paraguay proposed to end the controversy over ownership of the Salto del Guairá by building a binational hydroelectric dam that would flood them. The act was the definitive sign that a new alignment toward Brazil was taking place within Paraguay and that Brazil was extending its influence westward, into a zone previously under the dominance of Argentina. The destabilization of the balance of power led to the signing of 1969 Treaty of the Cuenca de la Plata between Argentina, Bolivia, Brazil, Paraguay, and Uruguay, unusual in its scope and the fact that it was not preceded by military engagement. When the Itaipú Treaty was signed in 1973 by Brazil and Paraguay, the Yacyretá Treaty between Argentina and Paraguay, languishing for decades though it had been proposed as far back as the 1920s, was suddenly drafted and signed. The positive experience of the successful negotiation of the tripartite 1979 Corpus-Itaipú Accord (Argentina, Brazil, Paraguay) resulted in another Act of Foz do Iguaçu (Argentina-Brazil, 1985) which called for a High Level Joint Commission for Integration between Argentina and Brazil, the first of several Argentina-Brazil agreements in the 1980s that heralded more than just cooperation but integration across a bellicose imperial boundary dating to the Treaty of Tordesillas (1494). The High Level Joint Commission planned a 1986 Program for Argentine-Brazilian Integration that was superseded in 1989 by the Integration, Cooperation, and Development Treaty (Argentina-Brazil). Up until this point, the Argentine-Brazilian closeness depended on the executive branch; treaties rely on that and the legislative branch, signaling that the proposed integration was increasingly important to both countries. The 1991 Treaty of Asunción followed, a push to foster economic integration along not merely bilateral, but multilateral lines. Signed by Argentina, Brazil, Paraguay, and Uruguay, the 1991 treaty founded Mercosur.

6. 1969 Plate Basin Treaty (Argentina, Bolivia, Brazil, Paraguay, Uruguay.) See Jessica Cattelino's (2008) work on watershed governance in the Everglades.

7. See Nikhil Anand's *Hydraulic City* (2017) for how water pressure within Mumbai's pipes, not just the presence of water infrastructure in that city's precarious neighborhoods, constitutes citizenship.

8. Böhmer 2009.

9. The discourse on human rights departs from earlier arrangements of rights and responsibilities, in spite of opposition from the ruling elite. By forcing space for change into the social-political pact, it has created an opportunity for the first time (in spite of multiple new constitutions in the past) for a new configuration of power among the citizenry. For a point of comparison, the foundation of a democracy in France is "common will"; in South Africa it is "common dignity."

10. Arendt 1976.

11. Anghie 1999, 2005; Folch 2016; Williams 1990.

12. Anghie 2005:251.

13. Leverett 2007.

14. Leverett 2007.

15. UNASUR was the product of a left-turn continental vision spearheaded by Brazilian president Lula da Silva, Argentine president Nestor Kirchner, and Venezuelan president Hugo Chavez. With shift beyond left-turn politics, the union has become less popular and may even evanesce.

16. UNASUR-OLADE 2012.

17. UNASUR-OLADE 2012:9.

18. UNASUR-OLADE 2012:9.

19. N. Smith 1993.

Conclusion

1. In technical terms, the treaty will be "revised."

2. Itaipu Binacional 2010:17.

3. Itaipu Binacional 2017:25.

4. Secretaria Técnica de Planificación de Desarrollo Económico y Social 2014:56. Belt et al. 2011:15.

5. As part of the Yacyretá renegotiations, Argentine president Mauricio Macri and Paraguayan president Horacio Cartes also signed an agreement to construct additional turbines in the Argentinian-Paraguayan binational dam on an outlet known as Aña Cua (which lies fully on the Paraguay side of the river). All the electricity from those new turbines would be contractually destined to the Argentinian market. It is unclear when construction on these new turbines will commence, and, given the delays attendant with Yacyretá's original construction, it is unclear whether this energy will be available by the time Paraguay's demand has reached its installed capacity.

6. Energy Information Administration 2017b: tables 1.3, 1.4a, 1.4b, and 2.1–2.6.

REFERENCES

Archives Consulted

ARGENTINA

El Litoral, Hemeroteca Digital, http://www.santafe.gov.ar/hemerotecadigital/articulo/ellitoral/.

BRAZIL

Veja, Acervo Digital, https://acervo.veja.abril.com.br/.
Folha de São Paulo, Acervo Digital, https://acervo.folha.com.br/index.do.
Brasil Nunca Mais, Digital, http://bnmdigital.mpf.mp.br/pt-br/.
Hemeroteca Digital Brasileira (Fundacao Biblioteca Nacional), http://bndigital.bn.gov.br/hemeroteca-digital/.
Jornal do Brasil, Digital, http://memoria.bn.br/hdb/periodo.aspx.

PARAGUAY

MCDyA (Museo y Centro de Documentación y Archivo para la Defensa de Derechos Humanos, Asunción, Paraguay).
1977 Carta abierta que dirigiera el Centro de Estudiantes de Ingeniería a cada Diputado Nacional en Junio de 1973. 00149F0351-0363.

Treaties and International Agreements

1872 Peace and Perpetual Friendship Treaty (Tratado de Paz y Amistad Perpetua entre Brasil y Paraguay) (Brazil-Paraguay).
1966 Act of Foz do Iguaçu (Acta de Foz do Iguaçu) (Brazil-Paraguay).
1969 Plate Basin Treaty (Tratado de la Cuenca de la Plata) (Argentina, Bolivia, Brazil, Paraguay, Uruguay).
1973 Itaipú Treaty (Tratado de Itaipú) (Brazil-Paraguay).
1973 Yacyretá Treaty (Tratado de Yacyretá) (Argentina-Paraguay).
1979 Tripartite Agreement for the Hydroelectric Use of Itaipú and Corpus (Acuerdo Tripartito para el Aprovechamiento Hidroeléctrico Itaipú y Corpus (Argentina-Brazil-Paraguay).
1985 Act of Foz do Iguaçu (Acta de Foz do Iguaçu) (Argentina-Brazil).
1991 Treaty of Asunción (Tratado de Asunción) (Argentina-Brazil-Paraguay-Uruguay).
2009 Joint Declaration (Declaración Conjunta) (Brazil-Paraguay).

Books and Articles

ABC Color

2006 Paraguay no quiere migajas, sino lo que le pertencece, advierte monseñor Lugo. http://archivo.abc.com.py/especiales/elecciones2008/articulos.php?fec=2006-12-05&pid=296614. Accessed July 17, 2014.

2008 Eletrobrás se presenta como la dueña absoluta de Itaipú. http://www.abc.com.py/2008-11-11/articulos/468557/eletrobras-se-presenta-como-la-duena-absoluta-de-itaipu. Accessed November 11, 2008.

2009a El fin de la Guerra Grande. July 24. http://www.abc.com.py/edicion-impresa/politica/el-fin-de-la-guerra-grande-4780.html. Accessed December 11, 2014.

2009b El acuerdo con Brasil iejo hacia la soberanía energética. July 26. http://www.abc.com.py/edicion-impresa/politica/el-acuerdo-con-brasil-avanza-hacia—la-soberania-energetica-5389.html. Accessed December 13, 2014.

2009c Acuerdos sobre Itaipú cambiarán tratado, dicen. October 6. Por Juan Antonio Franco. http://www.abc.com.py/edicion-impresa/economia/acuerdos-sobre-itaipu-cambiaran—tratado-dicen-27937.html. Accessed December 30, 2014.

2017 Brasil no puede prescindir de Itaipú en 2023, pero tendrá más opciones. http://www.abc.com.py/edicion-impresa/economia/brasil-no-puede-prescindir-de-itaipu-en-2023-pero-tendra-mas-opciones-1582561.html. Accessed September, 2017.

Abrams, Philip

1988 Notes on the Difficulty of Studying the State (1977). *Journal of Historical Sociology* 1 (1): 58–89.

Academia Brasileira de Ciências

2015 ABC promove coletiva de imprensa sobre a crise hídrica. http://www.abc.org.br/article.php3?id_article=4019. Accessed September 14, 2017.

Agamben, Giorgio

1998 *Homo Sacer: Sovereign Power and Bare Life*. Stanford, CA: Stanford University Press.

2005 *State of Exception*. Translated by Kevin Attell. Chicago: Chicago University Press.

Agencia Senado

2009 Delcídio destaca pontos do acordo entre Brasil e Paraguai sobre Itaipu. http://www12.senado.gov.br/noticias/materias/2009/11/03/delcidio-destaca-pontos-do-acordo-entre-brasil-e-paraguai-sobre-itaipu, November 11. Accessed December 26, 2014.

Alcaraz Gavilán, Pablo Ramón

2014 Influencia de los royalties y compensaciones generadas por las Entidades Binacionales Itaipú y Yacyretá, en el desarrollo humano y la pobreza extrema de la población paraguaya, en el período 1.989–2.010. Graduation thesis. Universidad Americana, Asunción.

Amery, Hussein A.

2002 Water Wars in the Middle East: A Looming Threat. *Geographical Journal* 168 (4): 313–23.

Anand, Nikhil

2011 Pressure: The PoliTechnics of Water Supply in Mumbai. *Cultural Anthropology* 26 (4): 542–64.

2017 *Hydraulic City: Water and the Infrastructures of Citizenship in Mumbai*. Durham, NC: Duke University Press.

ANDE (Administración Nacional de Electricidad)

n.d. Historia. http://www.ande.gov.py/historia.php. Accessed July 1, 2018.

Anghie, Antony

1999 Finding the Peripheries: Sovereignty and Colonialism in Nineteenth-Century International Law. *Harvard International Law Journal* 40 (1): 1–80.

2005 *Imperialism, Sovereignty and the Making of International Law*. Cambridge: Cambridge University Press.

Appel, Hannah

2012 Offshore Work: Oil, Modularity, and the How of Capitalism in Equatorial Guinea. *American Ethnologist* 39 (4): 692–709.

Apter, Andrew

2005 *The Pan-African Nation: Oil and the Spectacle of Culture in Nigeria*. Chicago: University of Chicago Press.

Aravamudan, Srinivas

2013 The Catachronism of Climate Change. *Diacritics* 41 (3): 6–30.

Arendt, Hannah

1976 *The Origins of Totalitarianism*. New York: Harcourt Brace.

1990 *On Revolution*. London: Penguin.

Auty, Richard

1993 *Sustaining Development in Mineral Economies: The Resource Curse Thesis*. New York: Routledge.

Baer, Werner, and Paul Beckerman

1989 The Decline and Fall of Brazil's Plan Cruzado. *Latin American Research Review* 24 (1): 35–64.

Ballestero, Andrea

2015 The Ethics of a Formula: Calculating a Financial-Humanitarian Price for Water. *American Ethnologist* 42:262–78.

Barbosa, Rubens

2013 "No alvo do Paraguai: Em 2023, sem ter investido país ficar com metade de Itaipu." O Globo. http://oglobo.globo.com/opiniao/no-alvo-do-paraguai-9133553.

Barboza, Mario Gibson

1992 *Na Diplomacia, o Traço Todo da Vida*. Rio de Janeiro, RJ: Editora Record.

Bartolomé, Mariano César

2002 La Triple Frontera: Principal Foco de Inseguridad en el Cono Sur Americano. *Military Review* (Spanish version), July August, 61–74.

Bayart, Jean-Francois

2000 Africa in the World: A History of Extraversion. *African Affairs* 99 (395): 217–67.

BBC Mundo

2005 Forum. Participe: Triple Frontera ¿culpable o victim? March 24. http://news.bbc.co.uk/hi/spanish/specials/newsid_4361000/4361915.stm. Accessed September 14, 2017.

Belt, C., et al.

2011 *Situación de Energías Renovables en el Paraguay*. Deutsche Gesellschaft fúr Internationale Zusammenarbeit (GIZ) GmbH.

Benjamin, Walter

1986 *Reflections: Essays, Aphorisms, Autobiographical Writings*. Translated by Edmund Jephcott. New York: Schocken.

Betiol, Laercio

1983 *Itaipu. Modelo avançado de cooperação internacional na Bacia do Prata*. Rio de Janeiro: FGV, Instituto de Documentação, Editura de Fundação Getúlio Vargas.

Blanc, Jacob

2015 Enclaves of Inequality: Brasiguaios and the Transformation of the Brazil-Paraguay Borderlands. *Journal of Peasant Studies* 42 (1): 145–58.

2017 Land, Legitimacy, and Dictatorship: The Itaipu Dam and the Visibility of Rural Brazil. Duke University, Global Brazil Lab, February 17, 2017.

Blanco, Gerardo., et al.

2017 Energy Transitions and Emerging Economies: A Multi-criteria Analysis of Policy Options for Hydropower Surplus Utilization in Paraguay. *Energy Policy* 108 (September): 312–21.

Blanco Sánchez, Jesús L.

1968 El Paraguay Como Fuente Energética. El Kanendiyú Guasú o Gran Salto del Guayrá. Su aprovechimiento hidroeléctrico según los Proyectos Brasileños.

Böhmer, Martin

2009 Lecture on Human Rights, Fulbright Regional Enhancement Seminar, April 22. Buenos Aires.

Boyer, Dominic

2011 Energopolitics and the Anthropology of Energy. *Anthropology News* 52:5–7.

Brantlinger, Patrick

1996 *Fictions of State: Culture and Credit in Britain, 1694–1994.* Ithaca, NY: Cornell University Press.

Bubandt, Nils

2009 From the Enemy's Point of View: Violence, Empathy, and the Ethnography of Fakes. *Cultural Anthropology* 24 (3): 553–88.

Bulmer-Thomas, Victor

1994 *The Economic History of Latin America since Independence.* Cambridge: Cambridge University Press.

Búsqueda

2009 Con las represas en situación "catastrófica," UTE consumió un tercio de petróleo importado y compra energía cara a Brasil. 23 de Abril, p. 15.

CADEP (Centro de Análisis y Difusión de la Economía Paraguaya)

2010 Monitoreo Fiscal Número 4. Asunción: Arandurã Impresiones.

2012a Cartilla Fiscal 11. Asunción: Arandurã Impresiones.

2012b Monitoreo Fiscal Número 4. Asunción: Arandurã Impresiones.

2017 Monitorea Fiscal: Evolución de las Cuentas. Asunción: Arandurã Impresiones.

California Energy Commission

2016 *Tracking Progress: Statewide Energy Demand.* http://www.energy.ca.gov/renewables/tracking_progress/documents/statewide_energy_demand.pdf. Accessed June 8, 2017.

Campbell, Jeremy M.

2014 Speculative Accumulation: Property-Making in the Brazilian Amazon. *Journal of Latin American and Caribbean Anthropology* 19 (2): 237–59.

Canese, Ricardo

2007 [2006] *La recuperación de la soberanía hidroeléctrica del Paraguay: En el marco de Políticas de Estado de energía.* Asunción: Editorial El ombligo del mundo.

Caplan, Patricia

2000 *Risk Revisited.* London: Pluto.

Cardoso, Fernando Henrique, and Enzo Faletto

1979 *Dependency and Development in Latin America.* Translated by Marjory Mattingly Urquidi. Berkeley: University of California Press.

Cardoso, Fernando Henrique, and José Guilherme C. Magnani
1974 Las contradicciones del desarrollo asociado *Desarrollo Económico* 14 (53) (April–June): 3–32.
Casa América
2008 Carlos Mateo Balmelli video.
Castañeda, Jorge G.
2006 Latin America's Left Turn. *Foreign Affairs* 85 (3): 28–43.
Cattelino, Jessica
2008 *High Stakes: Florida Seminole Gaming and Sovereignty*. Durham, NC: Duke University Press.
Cernea, Michael M.
2004 Social Impacts and Social Risks in Hydropower Programs: Pre-emptive Planning and Counter-Risk Measures. Keynote address: Session on Social Aspects of Hydropower Development United Nations Symposium on Hydropower and Sustainable Development. Beijing, China, October 27–29, 2004.
Céspedes, Angel Rafael
1982 *Proceso Analitico para Salvaguardar la Soberania del Paraguay en Itaipú*. Asunción: n.p.
Chakrabarty, Dipesh
2009 The Climate of History: Four Theses. In *Critical Inquiry*, 197–222. Chicago: University of Chicago Press.
Chibnik, Michael
2011 *Anthropology, Economics, and Choice*. Austin: University of Texas Press.
Coase, Ronald
1937 The Nature of the Firm. *Economica* 4 (16): 386–405.
Cohen, Felix
1960 The Spanish Origins of Indian Rights in the Law of the United States. In *The Legal Conscience: Selected papers of Felix S. Cohen*, edited by Lucy Kramer Cohen, 230–52. New Haven, CT: Yale University Press.
Coleman, Leo
2017 *A Moral Technology: Electrification as Political Ritual in New Delhi*. Ithaca, NY: Cornell University Press.
Colistete, Renato
2010 Revisiting Import-Substituting Industrialization in Post-war Brazil. Paper. http://www.fea.usp.br/feaecon//media/livros/file_517.
Colson, Elizabeth, and University of Zambia, Institute for African Studies
1971 *The Social Consequences of Resettlement: The Impact of the Kariba Resettlement upon the Gwembe Tonga*. Kariba Studies 4. Manchester, UK: Published on behalf of the Institute for African Studies, University of Zambia by Manchester University Press.
Comaroff, Jean, and John Comaroff
2001 *Millennial Capitalism and the Culture of Neoliberalism*. Durham, NC: Duke University Press.
Contraloría General de la República, Paraguay (Office of the Comptroller)
2012 Segundo informe avance de auditoría. *Examen Especial*. Deuda de la Entidad Binacional Itaipú.
Coronil, Fernando
1997 *The Magical State: Nature, Money, and Modernity in Venezuela*. Chicago: University of Chicago Press.
Coronil, Fernando, and Julie Skurski, eds.
2006 *States of Violence*. Ann Arbor: University of Michigan Press.

Craide, Sabrina
2009 Lobão diz que Paraguai não contribuiu financeiramente na construção da usina de Itaipu. Agência Brasil. http://agenciabrasil.ebc.com.br/noticia/2009-03-13/lobao-diz-que-paraguai-nao-contribuiu-financeiramente-na-construcao-da-usina-de-itaipu. Accessed March 13, 2009.

Crutzen, Paul
2002 Geology of Mankind. *Nature* 415 (6867): 23.

Das, Veena, and Deborah Poole
2004 *Anthropology in the Margins of the State.* Santa Fe, NM: School of American Research Press.

Davenport, E., C. Folch, and C. Vasu
n.d. Itaipú Dam: Paraguay's Growth Potential. White paper. https://itaipupost2023.com/white-paper/.

Debernardi, Enzo
1996 *Apuntes para la Historia Política de Itaipú.* Asunción: Editorial Gráfica Continua S.A.

De la Cadena, Marisol
2015 *Earth Beings: Ecologies of Practice across Andean Worlds.* Durham, NC: Duke University Press.

Department of State (US)
1976 Memorandum of Conversation between Secretary of State Henry Kissinger and Argentine Foreign Minister Adm. Cesar Guzzetti, Secret, June 10. https://nsarchive2.gwu.edu/NSAEBB/NSAEBB514/docs/Doc%2004%20-%2024087%20108254%201.pdf. Accessed August 6, 2018.

Depetris, P. J.
2007 The Parana River under Extreme Flooding: A Hydrological and Hydro-geochemical Insight. *Interciencia* 32 (10): 656–62.

Dinges, John
2004 *The Condor Years: How Pinochet and His Allies Brought Terrorism to Three Continents.* New York: New Press.

Dorcey, Anthony, International Union for Conservation of Nature and Natural Resources, and World Bank Group
1997 Large Dams: Learning from the Past, Looking at the Future. Workshop proceedings, Gland, Switzerland, April 11–12, 1997. Gland, Switzerland: IUCN.

Douglas, Mary, and Aaron Wildavsky
1983 *Risk and Culture: An Essay on the Selection of Technological and Environmental Dangers.* Berkeley: University of California Press.

Drackl é, Dorle, and Werner Krauss
2011 Ethnographies of Wind and Power. *Anthropology News* 52:9.

El Litoral
1977 Cuestiones con el Paraguay tratáronse. http://www.santafe.gov.ar/hemerotecadigital/diario/32247/.

Empresa de Pesquisa Energética (Brasil)
2017 *Balanço Energético Nacional 2017: Ano base 2016/Empresa de Pesquisa Energética.* Rio de Janeiro: EPE.

Energy Information Administration (US, EIA)
n.d. Energy Explained. https://www.eia.gov/Energyexplained/?page=us_energy_transportation. Accessed March 30, 2017.
2009 Annual Review 2008. Washington DC: US Department of Energy.
2014 Brazil Country Analysis. Washington DC: US Department of Energy.

2017a International Data Washington DC: US Department of Energy. https://www.eia.gov/energyexplained/?page=us_energy_home. Accessed September 13, 2017.

2017b Monthly Energy Review, April. Tables 1.3, 1.4a, 1.4b, and 2.1–2.6. https://www.eia.gov/energyexplained/?page=us_energy_home. Accessed September 13, 2017.

Escobar, Arturo

2010 Latin America at a Crossroads. *Cultural Studies* 24 (1): 1–65.

Espínola, Nancy

2009 Lugo había prometido renegociar Itaipú y no aceptar más migajas. ABC Color. http://www.abc.com.py/articulos/lugo—habia-prometido-renegociar-itaipu—y-no-aceptar-mas-migajas-52918.html. Accessed July 17, 2014.

Evans, Peter, Dietrich Rueschemeyer, and Theda Skocpol

1985 *Bringing the State Back In*. Cambridge: Cambridge University Press.

Ferradás, Carmen

1998 *Power in the Southern Cone Borderlands: An Anthropology of Development Practice*. Westport, CT: Bergin and Garvey.

Fischer, Michael

2005 Technoscientific Intrastructures and Emergent Forms of Life: A Commentary. *American Anthropologist* 107 (1): 55–61.

Flecha, Agustin Oscar

1975 *Distribución de Ingreso y Subdesarrollo: Un Model Matemático; Impacto del Gasto de Inversión de Itaipu en la Economía Nacional*. Asunción: Escuela Técnica Salesiana.

Fogel, Ramón, and Marcial Riquelme, eds.

2005 *Enclave sojero: Merma de soberanía y pobreza*. Asunción: Centro de Estudios Rurales Interdisciplinarios.

Folch, Christine

2013 Surveillance and State Violence in Stroessner's Paraguay: Itaipú Hydroelectric Dam, Archive of Terror. *American Anthropologist* 115 (1): 44–57.

2015 The Cause of All Paraguayans? Defining and Defending Hydroelectric Sovereignty. *Journal of Latin American and Caribbean Anthropology* 20 (2): 242–63.

2016 The Nature of Sovereignty in the Anthropocene: Lessons of Struggle, Otherness, and Economics from Paraguay. *Current Anthropology* 57 (5): 565–85.

2018 Catastrophic Hypotheses, Nuclear Fears, and Hydroelectric Terrorism as State Power in South America. Paper presented at 117th Annual Meeting of the American Anthropological Association, San Jose, CA. November 16.

Folha de São Paulo

1977 Brasil toma decisão unilateral sobre ciclagem 11 do Novembre, pagina 26. http://acervo.folha.com.br/resultados/?q=itaipu&site=fsp&periodo=acervo&x=18&y=19.

Foucault, Michel

1991 Governmentality. In *The Foucault Effect: Studies in Governmentality*, ed. Graham Burchell, Colin Gordon, and Peter Miller, 87–104. Chicago: University of Chicago Press.

Galeano, Eduardo

1997 *Open Veins of Latin America: Five Centuries of the Pillage of a Continent*. New York: Monthly Review Press.

Gamón, Efraín Enríquez

2007 *Itaipú: Aguas que valen oro*. 2nd ed. Asunción: Editorial Gráfica Mercurio.

Garsten, Christina, and Anna Hasselström

2003 Risky Business: Discourses of Risk and (Ir)responsibility in Globalizing Markets. *Ethnos* 68 (2): 249–70.

González Echevarría, Roberto

1990 *Myth and Archive: A Theory of Latin American Narrative.* Cambridge: Cambridge University Press.

Graeber, David

2001 *Toward an Anthropological Theory of Value: The False Coin of our Own Dreams.* New York: Palgrave.

2011 *Debt: The First Five Thousand Years.* New York: Melville House.

Graham, Steve, and Simon Marvin

2001 *Splintering Urbanism: Networked Infrastructures, Technological Mobilities, and the Urban Condition.* New York: Routledge.

Gregory, Christopher

2012 On Money, Debt, and Morality: Some Reflections on the Contribution of Economic Anthropology. *Social Anthropology* 20 (4): 380–96.

Gualdani, Fernando

2008 Arranca la batalla eléctrica suramericana. *El País,* December 6:2.

Gudeman, Stephen

2001 *The Anthropology of Economy: Community, Market, and Culture.* Malden, MA: Blackwell.

Gudynas, Eduardo

2011 El nuevo extractivismo progresista en América del sur: Tesis sobre un iejo problema bajo nuevas expresiones. *Colonialismos del Siglo* 21:75–92. Icaria Editorial, Barcelona.

Gupta, Akhil

2015 An Anthropology of Electricity from the Global South. *Cultural Anthropology* 30(4): 555–68. https://doi.org/10.14506/ca30.4.04. Accessed January 2, 2019.

Hansen, Thomas B., and Finn Stepputat

2006 Sovereignty Revisited. *Annual Review of Anthropology* 35:295–315.

Harvey, David

2006 *Spaces of Global Capitalism: Towards a Theory of Uneven Geographical Development.* London: Verso.

2007 *A Brief History of Neoliberalism.* Oxford: Oxford University Press.

Harvey, Penelope

2005 The Materiality of State-Effects: An Ethnography of a Road in the Peruvian Andes. In *State Formation: Anthropological Explorations,* ed. C. Krohn-Hansen and K. Nustad, 216–47. Cambridge: Pluto.

2014 Infrastructures of the Frontier in Latin America. *Journal of Latin American and Caribbean Anthropology* 19 (2): 280–83.

Hausman, William J., and John L. Neuefeld

1997 The Rise and Fall of American and Foreign Power Company: A Lesson from the Past? *Electricity Journal* 10 (1): 46–53.

Hetherington, Kregg

2011 *Guerrilla Auditors: The Politics of Transparency in Neoliberal Paraguay.* Durham, NC: Duke University Press.

2014 Waiting for the Surveyor: Development Promises and the Temporality of Infrastructure. *Journal of Latin American and Caribbean Anthropology* 19 (2): 195–211.

Hetherington, Kregg, and Jeremy Campbell

2014 Nature, Infrastructure, and the State: Rethinking Development in Latin America. *Journal of Latin American and Caribbean Anthropology* 19 (2): 191–94.

Ho, Karen

2009 *Liquidated: An Ethnography of Wall Street.* Durham, NC: Duke University Press.

Howe, Cymene

2015 Latin America in the Anthropocene: Energy Transitions and Climate Change Mitigations. *Journal of Latin American and Caribbean Anthropology* 20:231–41.

Howe, Cymene, and Dominic Boyer

2015 Aeolian Politics. *Distinktion: Scandinavian Journal of Social Theory* 16:1–18.

Humphrey, Caroline

2005 Ideology in Infrastructure: Architecture and Soviet Imagination. *Journal of the Royal Anthropological Institute* 11 (1): 39–58.

Hudson, Rex

2003 Terrorist and Organized Crime Groups in the Tri-border Area (TBA) of South America. Report prepared by the Federal Research Division, Library of Congress. July.

Hulme, Michael

2016 The Cultural Functions of Climate. Keynote lecture, Anthropology, Weather and Climate Change. Royal Anthropological Institute, British Museum. May.

IACHR (Inter-American Commission on Human Rights)

1978 Informe sobre la situación de los derechos humanos en Paraguay. http://www.cidh.org/countryrep/Paraguay78sp/indice.htm. Accessed August 31, 2010.

Inter-American Development Bank

http://www.idb.int/projects/project.cfm?id=PR0002&lang=en. Accessed August 31, 2010.

IPCC (Intergovernmental Panel on Climate Change)

2014 Fifth Assessment Report.

Itaipu Binacional

n.d. Itaipu.gov.br/Itaipu.gov.py. Official website.

1994 Itaipu: Hydroelectric Project. Curitiba: Itaipu Binacional.

2003 Prestación de los Servicios de Electricidad y Bases Financieras. Compendio.

2006 Memoria Anual 2005.

2007 Memoria Anual 2006.

2008 Memoria Anual 2007.

2009a Memoria Anual 2008.

2009b Mateo Balmelli Brindó Conference en Curitiba. Itaipu Press Office. https://www.itaipu.gov.py/es/sala-de-prensa/noticia/mateo-balmelli-brindo-conferencia-en-curitiba. Accessed January 2, 2019.

2010 Memoria Anual 2009.

2011 Memoria Anual 2010.

2012 Memoria Anual 2011.

2013 Memoria Anual 2012.

2014 Memoria Anual 2013.

2015 Memoria Anual 2014.

2016 Memoria Anual 2015.

2017 Memoria Anual 2016.

2018 Memoria Anual 2017.

Johnston, Barbara Rose

2013 Human Needs and Environmental Rights to Water: a Biocultural Systems Approach to Hydrodevelopment and Management. *Ecosphere* 4 (3): 39.

Jornal do Brasil

1977 Argentina é metade das dores-de-cabeça de Cavalcanti. August 24, 17.

Joseph, Gilbert, and David Nugent, eds.

1994 *Everyday Forms of State-Formation: Revolution and the Negotiation of Rule in Modern Mexico*. Durham, NC: Duke University Press.

Karam, John Tofik
n.d. Manifold Destiny: Arabs at an American Crossroads. Unpublished book manuscript.
2006 *Another Arabesque: Syrian-Lebanese Ethnicity in Neoliberal Brazil.* Philadelphia: Temple University Press.

Karl, Terry Lynn
1997 *The Paradox of Plenty: Oil Booms and Petro-States.* Berkeley: University of California Press.

Kennedy, David
1986 Primitive Legal Scholarship. *Harvard International Law Journal* 27 (1): 1–98.

Kleinpenning, J.M.G., and E. B. Zoomers
1990 Colonización Interna y Desarrollo Rural: El Caso del Paraguay. *Revista Geográfica* 112:109–25.

Kraay, Hendrik, and Thomas L. Whigham
2005 *I Die with My Country: Perspectives on the Paraguayan War, 1864–1870.* Lincoln: University of Nebraska Press.

Krasner, Stephen
1999 *Sovereignty: Organized Hypocrisy.* Princeton, NJ: Princeton University Press.

Lambert, Peter
2006 "Muero con mi patria!" Myth, Political Violence and the Construction of National Identity in Paraguay. In *The Construction of National Identity and Political Violence in Latin America,* ed. W. Fowler and P. Lambert, 187–206. New York: Palgrave.

Larkin, Brian
2008 *Signal and Noise: Media, Infrastructure, and Urban Culture in Nigeria.* Durham, NC: Duke University Press.

Larson, Brooke
2004 *Trials of Nation Making: Liberalism, Race, and Ethnicity in the Andes, 1810–1910.* Cambridge: Cambridge University Press.

Las Casas, Bartolomé de
2007 *Historia de Las Indias.* Libro Segundo. Tomo 3. Alicante: Biblioteca Virtual Miguel de Cervantes. http://www.cervantesvirtual.com/obra-visor/historia-de-las-indias-tomo-3—0/html/. Accessed July 26, 2009.

Latour, Bruno
1986 [1979] *Laboratory Life: The Construction of Scientific Facts.* 2nd ed. Princeton, NJ: Princeton University Press.

Lépinay, Vincent Antonin
2011 *Codes of Finance: Engineering Derivatives in a Global Bank.* Princeton, NJ: Princeton University Press.

Leverett, Flynt
2007 The Geopolitics of Oil and American's International Standing. Testimony before the Committee on Energy and Natural Resources, US Senate. January 10. https://www.hsdl.org/?view&did=236360. Accessed January 2, 2019.

Limbert, Mandana
2010 *In the Time of Oil: Piety, Memory, and Social Life in an Omani Town.* Stanford, CA: University Press.

Lins Ribeiro, Gustavo
1994 *Transnational Capitalism and Hydropolitics in Argentina: The Yacyretá High Dam.* Gainesville: University Press of Florida.

Love, Thomas, and Cindy Isenhour
2016 Energy and Economy: Recognizing High-Energy Modernity as a Historical Period. *Economic Anthropology* 3:6–16.

Mains, Daniel

2012 Blackouts and Progress: Privatization, Infrastructure, and a Developmentalist State in Jimma, Ethiopia. *Cultural Anthropology* 27 (1): 3–27.

Mallon, Florencia E.

1995 *Peasant and Nation: The Making of Postcolonial Mexico and Peru.* Berkeley: University California Press.

Marengo, J. A., S. C. Chou, G. Kay, L. M. Alves, J. F. Pesquero, W. R. Soares, and P. Tavares

2012 Development of Regional Future Climate Change Scenarios in South America Using the eta CPTEC/HadCM3 Climate Change Projections: Climatology and Regional Analyses for the Amazon, São Francisco and the Parana River Basins. *Climate Dynamics* 38 (9–10): 1829–48.

Marx, Karl

1993 *Capital: A Critique of Political Economy.* Vol. 3. New York: Penguin Classics.

Masco, Joseph

2006 *The Nuclear Borderlands: The Manhattan Project in Post–Cold War New Mexico.* Princeton, NJ: Princeton University Press.

2014 *The Theater of Operations: National Security Affect from the Cold War to the War on Terror.* Durham, NC: Duke University Press.

Mateo Balmelli, Carlos

2011 *Itaipú: Una reflexión ético-político sobre el poder.* Asunción: ServiLibro.

Maurer, Bill

2012 Late to the Party: Debt and Data. *Social Anthropology* 20 (4): 474–81.

Mauss, Marcel

2000 *The Gift: The Form and Reason for Exchange in Archaic Societies.* New York: W. W. Norton.

Máximo, Luciano

2009 Paraguai faz ofensiva para vender energia. *Valor Econômico.* October 6, A5.

McCandless, Matthew

2007 Community Involvement in the Development of Small Hydro in Uttaranchal, India. Master's thesis, University of Manitoba.

Mignolo, Walter

2005 *The Idea of Latin America.* Malden: Blackwell.

Ministerio Coordinador de Sectores Estratégicos

n.d. Balance Energético Nacional 2016 (Ecuador).

Mintz, Sidney

1986 *Sweetness and Power: The Place of Sugar in Modern History.* New York: Penguin Books.

Mitchell, Timothy

1999 State, Economy, and the State Effect. In *State/Culture: State-Formation after the Cultural Turn,* ed. G. Steinmetz, 76–97. Ithaca, NY: Cornell University Press.

2011 *Carbon Democracy: Political Power in the Age of Oil.* New York: Verso.

Montero, Humberto

2008 Brasil debe ver si nos quiere como socios o dominarnos. *La Razón,* December 7, 14–16.

Morgan, Lewis Henry

1877 *Ancient Society; or, Researches in the Lines of Human Progress from Savagery through Barbarism to Civilization.* Chicago: C. H. Kerr.

Mrázek, Rudolf

2002 *Engineers of Happy Land: Technology and Nationalism in a Colony.* Princeton, NJ: Princeton University Press.

Mukerji, Chandra
2009 *Impossible Engineering: Technology and Territoriality on the Canal du Midi*. Princeton, NJ: Princeton University Press.

Nader, Laura
2010 *The Energy Reader*. Malden: Blackwell.

Nash, June C.
1993 *We Eat the Mines and the Mines Eat Us: Dependency and Exploitation in Bolivian Tin Mines*. New York: Columbia University Press.

Nickson, Andrew
1982 The Itaipu Hydro-electric Project: The Paraguayan Perspective. *Bulletin of Latin American Research* 2 (1): 1–20.

Nugent, David
1997 *Modernity at the Edge of Empire: State, Individual, and Nation in the Northern Peruvian Andes, 1885–1935*. Stanford, CA: Stanford University Press.

Ong, Aihwa
2006 *Neoliberalism as Exception: Mutations in Citizenship and Sovereignty*. Durham, NC: Duke University Press.

Organismo Supervisor de la Inversión en Energía y Minería
n.d. Evolución de generación de energía eléctrica por fuente de generación. Perú. http://observatorio.osinergmin.gob.pe/evolucion-por-fuente-de-generacion. Accessed July 13, 2018.

Orlove, Benjamin
2002 *Lines in the Water: Nature and Culture at Lake Titicaca*. Berkeley: University of California Press.

Parlasur
2008 Audiencia Pública Debate Sobre la Situación de Itaipu Binacional (29 de noviembre de 2008). Versión taquigráfica realizada por Signos. Montevideo: Parlasur.

Pastore, Carlos
1949 *La lucha por la tierra en el Paraguay*. Montevideo: Editorial Antequera.

Peebles, Gustav
2011 *The Euro and Its Rivals: Currency and the Construction of a Transnational City*. Bloomington: Indiana University Press.

Peralta, José
2009 En el diferendo de Itaipú, Brasil define si quere ser 'lider' o busca 'dominación' hegemónica,' dijo delegado paraguayo en hidroeléctrica. *Búsqueda*, March 19, 4.

Petersen, Anneliese
2014 The Grand Inga Dam: Schistosomiasis as an Example of the Differential Valuation of Human Life in Hydroelectric Dam Construction. Unpublished term paper. Wheaton, IL: Wheaton College.

Plato
1967 Euthydemus. In *Plato in Twelve Volumes*, vol. 3, trans. W.R.M. Lamb, 373–505. Cambridge, MA: Harvard University Press.

Poff, Leroy N., David J. Allan, Margaret A. Palmer, David D. Hart, Brian D. Richter, Angela H. Arthington, Kevin H. Rogers, Judy L. Meyer, and Jack A. Stanford
2003 River Flows and Water wars: Emerging Science for Environmental Decision Making. *Frontiers in Ecology and the Environment* 1:298–306.

Prebisch, Raúl
1981 *Capitalismo periférico: Crisis y transformación*. México, D.F.: Fondo de Cultura Económica.

Rama, Angel
1996 *The Lettered City*. Translated by John Charles Chasteen. Durham, NC: Duke University Press.
Ramón Hernández, O. P.
1991 The Internationalization of Francisco de Vitoria and Domingo de Soto. *Fordham International Law Journal* 15:1031.
Roa Bastos, Augusto
1974 *Yo, el Supremo*. Buenos Aires: Siglo Veintiuno Argentina Editores.
Robertson, John Parish, and William Parish Robertson
1839 *Francia's Reign of Terror, Being the Continuation of Letters on Paraguay*. Vol. 3. London: John Murray.
Rogers, Douglas
2012 The Materiality of the Corporation: Oil, Gas, and Corporate Social Technologies in the Remaking of a Russian Region. *American Ethnologist* 39 (2): 284–96.
2014 Petrobarter: Oil, Inequality, and the Political Imagination in and after the Cold War. *Current Anthropology* 55 (2): 131–55.
Roitman, Janet
2004 *Fiscal Disobedience: An Anthropology of Economic Regulation in Central Africa*. Princeton, NJ: Princeton University Press.
Ross, Michael
1991 The Political Economy of the Resource Curse. *World Politics* 51 (1): 297–322.
Sarmiento, Domingo
1998 *Facundo; or, Civilization and Barbarism*. New York: Penguin Books.
Sassen, Saskia
1999 De-nationalization: Some Conceptual and Empirical Elements. *PoLAR* 22 (2): 1–16.
Sawyer, Suzana
2004 *Crude Chronicles: Indigenous Politics, Multinational Oil, and Neoliberalism in Ecuador*. Durham, NC: Duke University Press.
Schivelbusch, Wolfgang
1987 *The Railway Journey: Industrialization and Perception of Time and Space*. Berkeley: University of California Press.
Schmitt, Carl
1985 *Political Theology: Four Chapters on the Concept of Sovereignty*. Translated by George Schwab. Chicago: University of Chicago Press.
Schoen, Douglas, and Michael Rowan
2009 *The Threat Closer to Home: Hugo Chavez and the War against America*. New York: Simon and Schuster.
Schuster, Caroline
2015 *Social Collateral: Women and Microfinance in Paraguay's Smuggling Economy*. Berkeley: University of California Press.
Schwenkel, Christina
2015 Spectacular Infrastructure and Its Breakdown in Socialist Vietnam. *American Ethnologist* 42 (3): 520–34.
Scott, James Brown
1934 *The Spanish Origins of International Law*. Oxford: Clarendon.
Secretaria Técnica de Planificación de Desarrollo Económico y Social
2014 Plan Nacional de Desarrollo 2030. Asunción, Paraguay.
Sharma, Aradhana, and Akhil Gupta, eds.
2006 *Anthropology of the State*. Malden: Blackwell.

Smith, Adam

2014 *The Wealth of Nations*. London: Shine Classics.

Smith, Neil

1993 Homeless/Global: Scaling Places. In *Mapping the Futures: Local Cultures, Global Change*, ed. John Bird et al., 87–119. London: Routledge.

2002 New Globalism, New Urbanism: Gentrification as Global Urban Strategy. *Antipode*, 34 (3): 427–50.

Smith-Nonini, Sandy

2016 The Role of Corporate Oil and Energy Debt in Creating the Neoliberal Era. *Economic Anthropology* 3 (1): 57–67.

Song, Jesook

2009 *South Koreans in the Debt Crisis: The Creation of a Neoliberal Welfare Society*. Durham, NC: Duke University Press.

Star, Susan Leigh

1999 The Ethnography of Infrastructure. *American Behavioral Scientist* 43 (3): 377–91.

Strathern, Marilyn

1975 *No Money on Our Skins*. New Guinea Research Bulletin 61. Port Moresby: Australian National University.

Stroessner, Alfredo

1965 Discurso Pronunciado en el acto inaugural de la VIa. Reunión de la Asamblea de Gobernadores del Banco Interamericano de Desarrollo, 26 de Abril. Subsecretaria de informaciones y cultura de la presidencia de la república. Asunción, Paraguay.

Svampa, Maristella

2011 Extractivismo neodesarrollista y movimientos sociales: ¿Un giro ecoterritorial hacia nuevas alternativas? In *Más Allá Del Desarrollo*, 185–218. Grupo Permanente de Trabajo sobre Alternativas al Desarrollo. Miriam Lang y Dunia Mokrani, compilers. Quito, Ecuador: Editorial El Conejo.

Telesur

2016 Venezuela en emergencia eléctrica por efectors del Fenómeno El Niño. March 27. https://www.telesurtv.net/news/Emergencia-electrica-en-Venezuela-20160427-0005.html. Accessed July 13, 2018.

Tilly, Charles

1975 *The Formation of National States in Western Europe*. Princeton, NJ: Princeton University Press.

Toledano, Perrine, and Nicolas Maennling

2013 Leveraging Paraguay's Hydropower for Sustainable Economic Development, Consultation Draft. Vale Columbia Center on Sustainable International Investment.

Trouillot, Michel

2003 *Global Transformations: Anthropology and the Modern World*. London: Palgrave.

Tsing, Anna Lowenhaupt

2005 *Friction: An Ethnography of Global Connection*. Princeton, NJ: Princeton University Press.

Tulachan, Bikul M.

2008 Caste-Based Exclusion in Nepal's Communal Micro-hydro Plants. *Undergraduate Economic Review* 4 (1): article 14.

Última Hora

2009a Existe apoyo clave de Cardoso a interés paraguayo en Itaipú. http://www.ultimahora.com/existe-apoyo-clave-cardoso-interes-paraguayo-itaipu-n262084.html. Accessed December 30, 2014.

2009b Mateo: "Samek es mi amigo." February 19. http://www.ultimahora.com/notas/197701-Mateo-Samek-es-mi-amigo. Accessed February 19, 2009.

UNASUR-OLADE

2012 UNASUR: Un Espacio que Consolida la Integración Energética. Quito, Ecuador.

Unidad de Planificación Minero Energética

n.d. Energía Eléctrica SIN (Sistema Interconectado Nacional). Colombia. http://www1.upme.gov.co/InformacionCifras/Paginas/PETROLEO.aspx. Accessed July 13, 2018.

Ventura, Manoel

2018 Fase de grandes hidrelétricas chega ao fim: Com privatização da Eletrobras e restrição ambiental, pequenas usinas, energia eólica e solar devem ganhar espaço. *O Globo*, February 1. https://oglobo.globo.com/economia/fase-de-grandes-hidreletricas-chega-ao-fim-22245669. Accessed June 23, 2018.

Viceministerio de Minas y Energía (Paraguay)

2016 Balance Energético Nacional. http://www.ssme.gov.py/vmme/pdf/balance2015/Balance%20Energetico%20Nacional%202015.pdf. Accessed June 24, 2017.

Vitoria, Francisco de

1917 [1532] *De Indis et de Iure Belli Relectiones: Being Parts of Relectiones Theologicae XII.* Edited by Ernest Nys. Translated by John Pawley Bate. Washington, DC: Carnegie Institution.

Viveiros de Castro, Eduardo

1998 Cosmological Deixis and Amerindian Perspectivism. *Journal of the Royal Anthropological Institute* 4 (3): 469–88.

von Schnitzler, Antina

2008 Citizenship Prepaid: Water, Calculability, and Techno-politics in South Africa. *Journal of South African Studies* 34 (4): 899–917.

2013 Traveling Technologies: Infrastructure, Ethical Regimes, and the Materiality of Politics in South Africa. *Cultural Anthropology* 28 (4): 670–93.

Warth, Anne and Tânia Monteiro

2019 Governo quer rever acordo que faz Brasil pagar mais que Paraguai pela energia de Itaipu. *Estadão*, February 19. https://economia.estadao.com.br/noticias/geral,governo-quer-rever-acordo-que-faz-brasil-pagar-mais-que-paraguai-pela-energia-de-itaipu,70002728224. Accessed February 19, 2019.

Wasmosy, Juan Carlos

2008 *Archivo Itaipú: Memorias y documentos inéditos.* Asunción: ColorShop Estación Gráfica SRL.

Watts, Michael

2008 Sweet and Sour. Working Paper No. 18. Niger Delta Economies of Violence Working Papers. Institute of International Studies, University of California, Berkeley.

Weber, Max

2004 *The Vocation Lectures: Politics as a Vocation.* Indianapolis: Hackett.

Weisskoff, Richard

1992 The Paraguayan Agro-export Model of Development. *World Development* 20 (10): 1531–40.

Weyland, Kurt

2003 Neopopulism and Neoliberalism in Latin America: How Much Affinity? *Third World Quarterly* 24 (6): 1095–115.

White, John Howard

2010 Itaipú: Gender, Community, and Work in the Alto Paraná Borderlands, Brazil and Paraguay, 1954–1989. PhD diss., University of New Mexico.

White, Leslie A.
1943 Energy and the Evolution of Culture. *American Anthropologist* 45:335–56.

White, Richard
1996 *The Organic Machine*. New York: Hill and Wang.

Wilkis, Ariel
2018 *The Moral Power of Money*. Palo Alto, CA: Stanford University Press.

Williams, Robert A.
1990 *The American Indian in Western Legal Thought: The Discourses of Conquest*. New York: Oxford University Press.

Winichakul, Thongchai
1997 *Siam Mapped: A History of the Geo-body of a Nation*. Honolulu: University of Hawaii Press.

Winther, Tanja, and Wilhite, Harold
2015 Tentacles of Modernity: Why Electricity Needs Anthropology. *Cultural Anthropology* 30 (4): 569–77.

Wittfogel, Karl
1957 *Oriental Despotism: A Comparative Study of Total Power*. New Haven, CT: Yale University Press.

Wolf, Aaron
1999 "Water Wars" and Water Reality: Conflict and Cooperation along International Waterways. In *Environmental Change, Adaptation, and Security*, ed. S. Lonergan, 251–65. NATO ASI Series 65. Dordrecht, The Netherlands: Kluwer Academic Press.

Wolf, Eric
1982 *Europe and the People without History*. Berkeley: University of California Press.

Wood, Caura
2016 Inside the Halo Zone: Geology, Finance, and the Corporate Performance of Profit in a Deep Tight Oil Formation. *Economic Anthropology* 3 (1): 43–56.

World Bank
n.d.a Brazil, Country Report. https://data.worldbank.org/country/brazil. Accessed April 3, 2017.

n.d.b Paraguay, Country Report. https://data.worldbank.org/country/paraguay. Accessed April 3, 2017.

2010 Paraguay Poverty Assessment: Determinants and Challenges for Poverty Reduction. Report No. 58638-PY. http://siteresources.worldbank.org/EXTLACREGTOPPOVANA/Resources/PY_Estudio_de_Pobreza_ingles.pdf. Accessed April 3, 2017.

INDEX

Note: Illustrations are indicated with *italic* page numbers.

A NOTE ON THE TYPE

This book has been composed in Adobe Text and Gotham. Adobe Text, designed by Robert Slimbach for Adobe, bridges the gap between fifteenth- and sixteenth-century calligraphic and eighteenth-century Modern styles. Gotham, inspired by New York street signs, was designed by Tobias Frere-Jones for Hoefler & Co.